Thomas Rawson Birks

Modern Physical Fatalism and the Doctrine of Evolution

Including an examination of H. Spencer's First principles. Second Edition

Thomas Rawson Birks

Modern Physical Fatalism and the Doctrine of Evolution
Including an examination of H. Spencer's First principles. Second Edition

ISBN/EAN: 9783337312411

Printed in Europe, USA, Canada, Australia, Japan

Cover: Foto ©berggeist007 / pixelio.de

More available books at **www.hansebooks.com**

MODERN PHYSICAL FATALISM

AND THE

DOCTRINE · OF EVOLUTION.

MODERN PHYSICAL FATALISM

AND THE

DOCTRINE OF EVOLUTION,

INCLUDING

AN EXAMINATION

OF

Mr H. SPENCER'S FIRST PRINCIPLES.

BY

THOMAS RAWSON BIRKS, M.A.
PROFESSOR OF MORAL PHILOSOPHY, CAMBRIDGE.

SECOND EDITION.

WITH A PREFACE IN REPLY TO THE STRICTURES OF
Mr H. SPENCER.

BY

C. PRITCHARD, D.D., F.R.S.
SAVILIAN PROFESSOR OF ASTRONOMY IN THE UNIVERSITY OF OXFORD, AND
LATE FELLOW OF ST JOHN'S COLLEGE, CAMBRIDGE.

London:
MACMILLAN AND CO.
1882

CONTENTS.

	PAGE
PREFACE TO THE SECOND EDITION, BY DR PRITCHARD	vii
PREFACE TO THE FIRST EDITION	xxiii
INTRODUCTION	1

CHAPTER I.
THE DOCTRINE OF THE UNKNOWABLE 5

CHAPTER II.
ON ULTIMATE IDEAS IN PHYSICS 30

CHAPTER III.
ON THE RELATIVITY OF KNOWLEDGE 54

CHAPTER IV.
THE RELATIVITY OF KNOWLEDGE ACCORDING TO SIR W. HAMILTON AND MR MILL 79

CHAPTER V.
ON THE REALITY OF MATTER 107

CHAPTER VI.

PAGE

THE INDESTRUCTIBILITY OF MATTER 128

CHAPTER VII.

§ I. THE CONTINUITY OF MOTION 150
§ II. THE CONSERVATION OF FORCE. 159

CHAPTER VIII.

THE TRANSFORMATION OF FORCE AND MOTION 174

CHAPTER IX.

ON LAWS OF ATTRACTION AND REPULSION 197

CHAPTER X.

ON CHOICE AND WILL IN PHYSICAL LAWS 217

CHAPTER XI.

ON EVOLUTION 239

CHAPTER XII.

ON HETEROGENEITY 261

CHAPTER XIII.

ON FORCE AND LIFE 272

CHAPTER XIV.

ON NATURAL SELECTION 290

PREFACE TO THE SECOND EDITION.

IN July 1880, four years after the publication of the present work, Mr SPENCER published an Appendix to one of the stereotyped editions of his *First Principles*, animadverting on certain passages written by Professor BIRKS, which in the opinion of his friends requires notice. The author of *Modern Physical Fatalism* had been laid low by sickness in the previous April, and has never been able even to know of the attack made upon him; application therefore was made to me, to defend a man, who is from circumstances beyond control, unable to defend himself. In so doing a strong injunction was imposed upon me, to confine my remarks carefully and strictly within the bounds of Christian courtesy and moderation. Nor was the injunction without its meaning, seeing that Mr Spencer assails his literary antagonist in language which possibly might provoke retaliation, and which must certainly leave much to regret on Mr Spencer's part. For instance, Mr Spencer has thought it becoming to represent the Author of *Modern Physical Fatalism* under the undignified figure of a child[1] pulling about and entangling a skein of silk for half an hour. He intimates that "an intricate plexus

[1] Appendix, p. 580, l. 15, &c.

of misrepresentations, misunderstandings, and perversions, fills the three hundred and odd pages forming the volume[1]." In another passage[2] he forgets himself so far as to say that " Professor Birks apparently thinks that, moved by the high motive of 'doing God service,' he is warranted in taking a course" the opposite of generous; and "that he would fail of his duty did any regard for generous dealing prevent him from making a point against an opponent of his creed." It seems to me difficult to understand how harsh and contemptuous and undignified language such as this, is likely to further the satisfactory conclusion of an important argument.

In order to substantiate these severe, but as I shall shew, wholly unmerited charges against our Professor's intellectual acumen and moral rectitude of dealing, Mr Spencer selects some seven or eight passages taken from *Modern Physical Fatalism*, and sets them forth as samples of the entire work, and as illustrative of the writer's intellectual and moral capacity. I propose to take these passages *seriatim*; and after a few words of comment or explanation, I shall leave them and Mr Spencer's criticisms thereon, to the judgment of the reader. My object, be it clearly understood, is neither to attack Mr Spencer's *First Principles*, nor to defend Professor Birks's *Physical Fatalism*, taken as wholes, but solely to enquire how far Mr Spencer is justified in applying the undignified and contemptuous language which he has adopted, to the several passages in question.

[1] Appendix, page 581, l. 22, &c.
[2] Appendix, page 584, l. 32, &c.

I.

Mr Spencer commences his remarks as follows[1]: That abundant warrant for this assertion, (*viz.* the assertion regarding the babyish entanglement of the skein of silk) is furnished by one of the earliest paragraphs in the Professor's book, in which the author represents Mr Spencer as saying that: "*Ultimate religious ideas cannot be conceived.*" "To conceive," Mr Spencer says, is to frame in thought; and as every idea is framed in thought, it is nonsense to say of any idea, that it cannot be conceived; "nonsense which I have nowhere uttered." Again, "My statement is, that ultimate scientific ideas are all representations of realities that cannot be comprehended; and the like is alleged of ultimate religious ideas." "The things which I say cannot be comprehended or conceived, are *not the ideas*, but the *realities* for which they stand."

No doubt, our Professor would have written more cautiously and accurately, had he quoted Mr Spencer's actual words at length. But surely Professor Birks is as much justified in saying that an idea is inconceivable, as he would be in saying that the picture of any given object is unpaintable. The entity, the reality, which is to be painted or idealized, Mr Spencer admits to be incomprehensible; is it wrong then, is it unintelligible, is it *nonsense*, to say, that the picture representing the entity in question cannot be painted? Any author, any philosopher might safely and properly write without danger of being misunderstood: "The picture of an incomprehensible object

[1] Appendix, p. 580.

cannot be painted: the idea thereof is inconceivable." Surely Professor Birks might so write, without laying himself open to the charge, that the adoption of such language is like an infant pulling about a skein of silk for half-an-hour. The contemptuous simile is here wholly inapplicable to the author of *Physical Fatalism*, and very unworthy of the author of *First Principles*.

These remarks are a sufficient reply to Mr Spencer's contemptuous criticism. But they are by no means all that justice requires to be said upon the subject. For, strange to say, Mr Spencer himself falls into the very same error, if indeed error it be, of saying that an *idea* is *inconceivable*, for which he so severely chastises Professor Birks; the only difference herein between the two writers being that Mr Spencer actually takes pains to emphasize the phrase. For, in page 36 of *First Principles*, Mr Spencer thus writes: "As unlimited duration is inconceivable, *all those formal ideas* into which it enters are *inconceivable*, and indeed, if such an expression is allowable, are the *more inconceivable* in proportion as the other elements of the idea are indefinite."

II.

"Further," Mr Spencer adds, "at the end of the first paragraph which deals with me, I am represented as teaching that 'religion is equivalent to nescience or ignorance, alone'." Mr Spencer, not without some vehemence, denies that he has ever so taught. He adds: "Though I hold that an Ultimate Being, known with absolute certainty as existing, but of whose nature we are in ignorance, is the sphere of religious feeling, Professor Birks says, I hold that the ignorance alone is the sphere for religious feeling." No doubt an author has a right to claim for himself, his own interpretation of his own writings; or, at all events, to disclaim intentions which he is conscious are not his own; nevertheless a critic who puts a reasonable and probable construction upon an author's writings, is not fairly to be charged with literary dishonesty or intellectual incapacity, if such interpretation varies from the intention of the writer.

Now, Professor Birks in the course of his reading Mr Spencer's *First Principles*, would, among many other passages of a like import, read words such as the following. "Very likely there will ever remain a need to give shape to that indefinite sense of an Ultimate Existence, which forms the basis of our intelligence. We shall always be under the necessity of contemplating it, as some mode of being, that is of representing it to ourselves in some form of thought, however VAGUE; and we shall not err in doing this, so long as we treat every notion we thus frame, as *merely a symbol, utterly without resemblance to that for which it stands*[1]." If then the sphere for religious feeling is thus represented by Mr Spencer himself (as vague, and associated with a symbol utterly without resemblance to that for which it stands), can the critic who represents this sphere for the religious sentiment, as equivalent to a sphere of nescience or ignorance alone, be justly stigmatized as acting with literary dishonesty, or as resembling a child ravelling a skein of silk? The author of *First*

[1] *First Principles*, Ed. 3, p. 113, also see p. 106, &c.

Principles no doubt feels that his views on Religion are misinterpreted; but it was quite open to him to have stated the fact with emphasis, without the adoption of contemptuous language wholly unwarranted, and which can only recoil on its author.

III.

Mr Spencer then proceeds as follows. "When, in the first sixteen lines specifically treating of my views, these three cases occur, *it may be imagined what an intricate plexus of misrepresentations*, misunderstandings and perversions, fills the three hundred and odd pages forming the work. Especially may it be anticipated that the metaphysical discussions, occupying five chapters, are so confused that it is impossible to deal with them. I must limit myself to giving a sample or two from this part of the work: one of them *illustrating* Professor Birks's *critical fairness*, and the other *his philosophical capacity*[1]."

The specimen of 'critical fairness' referred to above, is the following: Mr MILL had made use of the phrase "*Permanent possibilities of sensation,*" in relation to the constitution of Matter. Professor Birks combats the phrase, and its consequences, at considerable length. In one part of the Professor's work this phrase of Mr Mill's, occurs in the midst of other paragraphs relating to views of Mr Spencer's, in such a manner that Mr Spencer complains that the obnoxious phrase would, by an ordinary reader, be imputed to himself, as expressing his own views; whereas

---

[1] Appendix, p. 581, l. 19, &c.

his own views are antagonistic, nay, entirely opposite to the views of Mr Mill. But how stands the case in reality? The obnoxious phrase *"permanent possibilities of sensation"* first occurs on page 103 of Professor Birks's work. The Professor there writes, with some emphasis, as follows: "The strangeness of the paradox reaches its height, when *Mr Mill* would replace matter, as common minds understand it, by *his new phrase* "Permanent possibilities of sensation." In page 112, the same phrase is repeated; but now it occurs in such a collocation, that Mr Spencer complains it might naturally, though most wrongly, be imputed to himself. Let it be most carefully noticed that nine pages before this collocation, the phrase had been emphatically referred to the authorship of Mr Mill. But, it is still more noticeable, that once again in page 126 Professor Birks again emphatically refers the authorship to *Mr Mill;* writing as follows:

"In contrast with this simple Realism, every form of philosophical Non-realism, whether of BERKELEY......*or of* Mill, who adds to Berkeley's view a new phrase and replaces Matter by *permanent possibilities of sensation*......" So that again and again, the obnoxious phrase is distinctly and emphatically referred to *Mr Mill* as its author. In the face of this careful appropriation of ownership to Mill, it seems impossible that any intelligent reader could, from Professor Birks's volume, refer that ownership to Mr Spencer. Mr Spencer has here fallen into a serious error. So far however from attributing this oversight of Mr Spencer's to wilfulness of intention on his part, there is one easier and more probable solution, namely in the suggestion that

Mr Spencer had not carefully read the context which precedes and follows the phrase in question. Now that Mr Spencer's error, and its possible source, are explained, it cannot be properly doubted that he will regret having thus wrongfully applied to Professor Birks so harsh and unmerited a phrase as "*literary misdemeanour*," and he will be glad to withdraw the contemptuous remark "from this sample of *critical truthfulness*, let us pass now to a sample of critical acumen[1]." Professor Birks has in fact been guilty of no *misdemeanour* whatever; and the implied charge of untruthfulness is an unwarranted accusation.

IV.

The *sample of critical acumen*, selected from our Professor's *Physical Fatalism* when impartially examined, will as little justify Mr. Spencer's remark, as the sample of critical truthfulness proves, on testing it, to be *a literary misdemeanour*. Professor Birks is charged with confusing the philosophical meaning of the term 'phenomenon,' with the ordinary and popular meaning of a mere *visible* appearance. He says: "*everywhere his* expressions and arguments make manifest the fact that Professor Birks thinks the meaning of phenomenon in metaphysical discussion, is no wider than that implied by something *visible*[2]." If this were truly the case, then I should admit the applicability to Professor Birks, of the contemptuous metaphor of the Baby and its pastime with the skein of silk. But what will at once strike an impartial reader, is

[1] Appendix, p. 582, l. 11.
[2] Appendix, p. 582, l. 41, &c.

the antecedent improbability that a classical scholar and a highly educated gentleman, to say nothing of a fellow of a most noble college, in a famous University, could by any means be led into so glaring a blunder. "Sounds, smells, tastes," says Mr Spencer, "are in his (our Professor's) view, not phenomena; nor are touches, pressures, tensions. And hence it results that when a pound of salt is dissolved in water, and it ceases to be visible, its existence, phenomenally considered, ends; its continued power of affecting our senses *by weight*, to the same extent as before the solution, not being considered as a phenomenal manifestation of its existence[1]." The reply that I shall proceed to give to this stringent but entirely groundless criticism of Mr Spencer on Professor Birks's work, is I conceive complete; and it is not a little surprising that the acute author of *First Principles* should himself have failed to perceive it. Professor Birks makes in reality no such blunder regarding the term *phenomenon*, as Mr Spencer so emphatically lays to his charge: on the contrary our Professor is quite as explicit in his expression of the meaning of 'Phenomenon' as is his severe, but mistaken critic. And, what is still more remarkable, is the fact, that the residual phenomenon of *continued weight* selected by Mr Spencer is as strongly insisted on by our Professor also, as it is by his critic. Let us listen to Professor Birks. "Matter," says he, "in the concrete, i.e. in those specialities of arrangement which names define, and our *senses* (mark the *plural*) recognize, is destroyed continually. A sheet of paper is burnt, a heap of gunpowder is fired, a

[1] Appendix, p. 582, l. 2 from the bottom.

drop of water is evaporated, a cloud is dispersed by the sunbeams; and the sheet of paper, the gunpowder, the drop, the cloud cease to exist. *The matter may survive in other forms*, but the form from which its name was derived, is destroyed, and is no more....Next, the substance, in these destructions of the *form* which determines our *sensations* (plu.) does not pass away, and is not destroyed, at least within the limits of our experience. Observation and experiment reveal constantly the survival of the matter itself, *as marked* by WEIGHT *in other forms*......[1]."

So that the very same phenomenon *of continued weight*, which Mr Spencer implicitly denies to have been observed by Professor Birks, is after all the very phenomenon selected by our Professor, to illustrate a *phenomenon* of matter surviving after other phenomena, such as visibility, for instance, hardness etc. have disappeared! Surely the most obvious solution of Mr Spencer's mistake herein, as respects the implied charge of want of *critical acumen*, on the part of his literary opponent, is a second time to be found in the hypothesis that the author of *First Principles* may not have read *Modern Physical Fatalism* with sufficient attention. It remains, however, for the reader to judge how far the charge of a literary *misdemeanour*, laid at the door of Professor Birks, in reality attaches to Mr Spencer himself.

V.

I come at last to the physical questions involved in Mr Spencer's rejoinder. Here we might naturally expect

---

[1] *Modern Physical Fatalism*, p. 130.

that we had escaped from metaphysical entanglements, and were landed on secure ground. Practically, and in this particular case, the ground is doubtful and slippery. Mr Spencer charges our Professor with stating that a certain *doctrine of Potential Energy*, held by Mr (now Sir William) Grove, and by Dr Tyndall is the same as that maintained in *First Principles*. The identity of these views Mr Spencer entirely disclaims, and with it (I presume) all the peculiar consequences which logically are derivable from it. One would naturally suppose that the issue is thus narrowed within a small compass. On investigation, I do not find it so. After giving all the attention within my power to Mr Spencer's explanations of his views on dynamics in general, and the Conservation of Energy in particular, I find myself unable to apprehend with clearness Mr Spencer's views on this branch of Physics. For half a century it has been my lot to study or to expound the writings of Newton and of his illustrious successors, but I fail to discover anything like a similarity between their investigations, and the dynamical theorems enunciated by Mr Spencer. Neither can I discover the stand-point from which he regards them. I find also that mathematicians of great eminence and ability, are equally at a loss with myself in this respect. Of course I cannot answer for Professor Birks in this matter; but it is quite possible that he may have misinterpreted some facts in Mr Spencer's writings on questions which require great precision and some mathematical acumen in their accurate enunciation; and, in so doing, he possibly may have unintentionally done Mr Spencer injustice. I fear that if

I involved myself in the argument respecting Mr Spencer's views relating to *Potential Energy*, and subjects cognate to it, I might be liable to the same misinterpretations of Mr Spencer's meaning, myself; and I might lay myself open to the same sort of severe and unmerited remark which Mr Spencer applies to the author of *Physical Fatalism*, viz. that "*there can be but one opinion respecting the honesty of making the assumption.*" All that I can say regarding this portion of Mr Spencer's rejoinder, is, that so far as I can understand the point at issue, I am unable to see how Professor Birks's *honesty* can reasonably be called in question. He or any other competent mathematician, may very possibly misunderstand some of Mr Spencer's dynamical propositions, but surely this misapprehension may be pardonable, probably it is unavoidable, and certainly it is fully within the category of what is honourable.

VI.

Finally I am arrived at the last specific charge laid at the door of Professor Birks. It is a very heavy charge: happily it is quite certain that Mr Spencer is mistaken in his implied supposition, that Professor Birks has in any degree wilfully misrepresented him. Surely the case ought to be free from all doubt, before the following clauses can be warrantably applied to any writer. "It is commonly thought nothing but fair that if he (an author) has made an error (I say this hypothetically, for in this case I have no error to acknowledge) he should be allowed the benefit of any correction he makes. Professor Birks, however, apparently thinks that moved by the high motive of 'doing God service' he is warranted in taking the opposite course —perhaps thinks, indeed, that he would fail of his duty, did any regard for generous dealing prevent him from making a point against an opponent of his creed[1]." This is very severe language. The reader will soon be able to judge for himself if it be merited.

The question at issue is that Professor Birks represents the author of *First Principles* as asserting "that gravitation is a necessary result of the Laws of Space." Now what Mr Spencer has actually asserted is "that *the Law* (of gravitation) is not simply an empirical one, but one deducible mathematically from the relations of space." The solution however of this controversy regarding the distinction between the two propositions, lies, I think, in a nutshell; and it is this. Mathematicians and other writers on Gravitational Astronomy are in the habit of using the generic term *gravitation*, sometimes in a very wide, and sometimes in a very restricted sense; sometimes in the sense of attraction only, and at other times, in relation only to the *law* of the attraction; sometimes it is molecular gravitation, at other times it is the resultant gravitation of masses of matter. The context always sufficiently defines the sense in which the term gravitation, in any particular instance, is intended to be understood. Newton himself in the Third Book of the *Principia* uses indiscriminately the same word gravitation, to express either the fact of the attraction of gravitating molecules, or the *Law* of that attraction. The context always clearly defines the meaning

[1] Appendix, p. 584.

of the generic term. So also with Professor Birks. He, like Newton, uses the term *gravitation* in its various senses, with more or less distinction, from page 222 to page 230 and onwards in his *Physical Fatalism*, in the midst of which pages the obnoxious passage occurs. I do not hesitate to say that in no case in which Professor Birks has used the term *gravitation*, is the intelligent reader (and to such readers alone is the question of any moment) left in any possible doubt, as to the sense which is intended. It would be simply ridiculous to suppose that an author of Mr Spencer's great eminence could hold that the attraction (or gravitation) of one particle of matter towards another, could possibly be a mathematical consequence of the mere relations of space. Independently of this, the context both preceding and following in *Physical Fatalism* would necessarily preclude all doubt on the subject, were it conceivable that such doubt could arise. Nevertheless it is here important to add that Professor Birks does, when occasion demands, most distinctly limit his remarks on Mr Spencer's objectionable Theory of Gravitation, to the *Law*, as distinguished from the other facts included in the general Term. What, for instance, can be more explicit than the following words in chapter x. of *Physical Fatalism?* "Mr Spencer in the first and second editions of his *First Principles* applied it (necessity) expressly to Newton's LAW of Gravitation. Physicists were obliged to assume this law, *because* it results from the necessary conditions of geometrical space. But in his third edition after fifteen years, the statement is silently withdrawn. Its historical falsehood, if not its theoretical absurdity, seems at last to have been detected by its author. But no open retractation is made[1].

If however Mr Spencer's vehement condemnation of *the ethics of the question* applies chiefly to Professor Birks quoting from Edition *two*, of the *First Principles*, passages withdrawn in Edition *three*, then it may be well not to overlook the fact, that Mr Spencer himself, in the midst of his vehement condemnation, admits that he "*has no error to acknowledge*" as to edition *two*, in relation to the matter regarding Gravitation, which our Professor quotes and controverts.

I have thus, with such judicial coldness as I could command, examined *seriatim* the several charges which Mr Spencer has brought against the author of *Physical Fatalism*. It has, however, not been without a feeling of amazement, that I read some of the unbecoming phrases hurled by Mr Spencer against his philosophical opponent. I cannot doubt that the impartial reader will sympathize with me in regarding some of these phrases as very undignified and wholly unwarranted, and sometimes couched in language which should have no place in the writings of any author laying claim to literary cultivation. I should, nevertheless, have felt much stronger indignation, and greater surprise, if it were not a matter of history that several other authors of eminence, who have ventured to controvert some of Mr Spencer's many speculative views, have also met with a very similar treatment at his hands. In fact, this mode of meeting a literary adversary seems to be chronic with Mr Spencer. Under any circumstances,

---

[1] *Modern Physical Fatalism*, p. 217.

the contemptuous treatment of a literary adversary without unquestionable cause, by a writer of acknowledged influence, amounts to a misguidance of readers less highly informed than himself, and is, constructively, a public wrong. It is but little to say of it, that such mode of argument is utterly unworthy and unbecoming in a philosopher, and simply recoils on the author who forgets his own dignity in adopting it.

C. PRITCHARD.

UNIVERSITY OBSERVATORY, OXFORD,
June 1, 1882.

PREFACE TO THE FIRST EDITION.

THE present work, containing the substance of a course of Lectures during the year 1875-6, is an examination and review of the modern Fatalistic Philosophy, and Doctrine of Evolution, as unfolded in Mr Herbert Spencer's 'First Principles,' the first in a series of works which have gained no small degree of reputation and influence. As I believe the views they advocate to be radically unsound, full of logical inconsistency and contradiction, and flatly opposed to the fundamental doctrines of Christianity, and even the very existence of Moral Science, I have felt that no task could fall more directly within the range of my appointed duties, as Knightbridge Professor, than the attempt to place in a clear light the thorough unsoundness of the basis upon which they rest. It is difficult, and almost impossible, to make an inquiry of this kind attractive to general readers. But I think that those who take the pains to read my strictures, and compare them with the statements of the work to which they are a reply, will find the effort repaid by a clearer apprehension of the topics in debate, and a full conviction that the old foundations of morality and Christian faith emerge with fresh evidence of their truth from all the floods and currents of unbelieving speculation in these last days.

TRINITY PARSONAGE, CAMBRIDGE,
Sept. 28, 1876.

INTRODUCTION.

MORAL Theology, as one main branch of universal science, involves three great truths. "There is one living and true God, the Maker of all things, visible and invisible." He is before all things, and by Him all things consist and endure. He is not only the Creator and the Preserver, but the Lawgiver and Moral Governor of the world. The universe is far higher and nobler than merely a vast machine. It includes not only lifeless matter, but living and sentient creatures. Among these it includes rational and intelligent beings, men and angels, endued with a power of choice, with reason and will, who can recognize a law of duty, know and love the great Creator, and either yield Him honour and due obedience, or rebel and disobey. The world's history thus comprises higher elements than the constancy of physical laws alone. It includes the ideas of righteousness and holiness, of sin, rebellion, and disobedience, of moral degradation and possible ruin, and again of moral recovery and redemption. It is a further main doctrine of this theology, that sin has entered into the world, and moral evil has widely prevailed, and that there is around us an actual scheme of Providence for the recovery of the lost and guilty to holiness, happiness, and immortal life. The course of the world's changes is thus one of deep mystery, from the entrance and long sway of moral evil among the creatures whom God has made. But while clouds and darkness are around the unseen Lawgiver, justice and judgment are the habitation of His throne. There may be a prolonged conflict of good and evil, but the final victory of Divine goodness is sure. There is no blind Fate, working on without aim or purpose, but the continual unfolding and fulfilment of the eternal counsel of a God of wisdom and infinite goodness. "Of him, and through him, and to him are all things, to whom be glory for ever and ever."

A wholly opposite view, borrowed from the heathen philosophy of Epicurus, has lately been revived among us, in connexion with the modern advances and discoveries in physical science. Its exact form varies in each of the writers by whom it is espoused and maintained. But its main principles are these. The First Cause, the Unseen Power of the universe, cannot be known at all, and therefore cannot reasonably be served, loved, honoured, or obeyed. Man is only a product and result of physical laws, some strange and peculiar condensation of atomic forces. The course of the world is a ceaseless evolution, implying no plan, choice, or will on the part of the unseen and unknown Power, and including no choice, will, or moral good or evil on the part of men, but only a fated cycle of inevitable changes, determined by fixed mechanical laws alone. In this way a primitive nebula, called sometimes a fire-mist, has developed into worlds, suns and planets and living things, and will probably return, after countless ages, to nebulous mist, confusion, and darkness again.

In the present work I propose to examine some of the

main principles of this Fatalistic Theory, especially as unfolded by Mr H. Spencer in his 'First Principles,' which comes first in order in a series of works, designed to compose a new and improved system of philosophy. The currency which it has gained, and the high praises which it has received from not a few eminent names, are reasons why those who believe it to be wholly fallacious and misleading in its reasoning, and completely antichristian in its whole spirit and tone, should submit its statements to a close and searching inquiry.

The work consists of two books, the first on the Unknowable, and the second on the Knowable. The former begins with an attempt to prove that all the ultimate religious ideas are contradictory, and incapable of being conceived; that Atheism, Pantheism, and Theism are three conceptions equally incapable of proof, and that "the Power which the universe manifests is utterly inscrutable." The ultimate scientific ideas, in their turn, are pronounced equally inconceivable, and full of apparent contradictions. As the result of the whole inquiry, Religion and Science are to be reconciled and harmonized by the axiom, that science includes the whole sphere of man's knowledge, that religion has for its proper field blind instincts or emotions, and that its distinction is to lie beyond all human experience, so that it is equivalent to Nescience or Ignorance alone.

The second book deals with the doctrine of the Knowable. Philosophy must be shut out from a great part of the domain which once was thought to belong to it. It must renounce the hope to learn anything, or know anything, as to the nature of the First Cause, or the Supreme Power. The domain left to it is that of Physical Science. It concerns itself with coexistences and sequences among phenomena, grouping them together, and thus attempts to rise higher and higher in the discovery of their actual laws. So long as truths are known only apart, they cannot be called philosophy. Science is partly unified knowledge, and Philosophy is knowledge completely unified.

I propose to examine carefully some of the main principles in this building of vast pretensions, which has gained for its author from a leading experimentalist, along with the works that are based upon it, the high title of the Apostle of the Understanding. How far these praises of the system are deserved will perhaps be seen more clearly at the close of the following examination.

CHAPTER I.

THE DOCTRINE OF THE UNKNOWABLE.

THE Physical Fatalism, of which the outlines are unfolded in Mr Spencer's First Principles, stands in diametrical contrast to the science of Moral Theology. The *a priori* truths of one scheme of thought are the fundamental falsehoods of the other. I shall now review in succession some of the main axioms of the new philosophy, in order to place in a clearer light the truths they contradict and oppose.

The first main part of the system, the foundation of all the rest, is the doctrine of the Unknowable. The problem proposed at the outset is to reconcile Religion with Science. And this is effected in a very simple, but a very startling way. Science is identified with Physics. Religion is made another name for simple Nescience. Theology is pronounced to be a futile attempt to transcend the appointed limits of human thought. It is a vain and pretentious claim to know the Unknowable.

A hundred and twenty pages are spent in unfolding this main axiom of the new philosophy. Atheism, Pantheism, and Theism are three rival attempts to explain the secret of the universe. But they are equal failures. The secret cannot be explained. Religious Nihilism is the only consistent and tenable theory. It is the negative creed, summed up in one maxim—"The deepest, widest, and most certain of all facts is this, that the Power which the universe manifests to us is utterly inscrutable."

The true doctrine, held by Christian divines and philosophers in every age, is the exact reverse. Science is no mere synonym for Material Physics. It includes, as taught by Bacon in the 'De Augmentis,' and by all men of deep thought, three main divisions, Physics, Humanity, and Theology; or the knowledge of Nature, of Man, and of God. The first of these has two main divisions, the science of lifeless matter, and of living but irrational things.

Science, then, stands in no real contrast to Humanity and Theology, which are two main parts of it, and the last is its noblest and highest portion. Its true contrasts are with nescience on one side, and omniscience on the other. It implies further a knowledge organized, and reduced to system, and not left in separate, unconnected fragments. Its tendency is to unity and completeness. Hence true Science cannot be opposed to Theology. Some real or pretended knowledge of the great First Cause is essential for the unity of all knowledge beside. Atheism is scientific anarchy. Pantheism and Theism are the false and true alternatives, towards one or other of which philosophy, in its striving after unity, must inevitably tend.

Theology is the highest and noblest science. Hence it needs for its attainment, not only the best and purest efforts of human thought, but special help and guidance from that God whom it seeks to know, the great Source of all true wisdom. And He has promised that they who seek earnestly this highest of all sciences, the knowledge of the Holy One, however various the degrees of their attainment, shall never seek in vain.

The doctrine of the Unknowable consists in the rejection and reversal of this great doctrine of the Christian Faith. It would reconcile Religion with Science by proving that they have nothing in common. Religion, in its creed, is another name for simple ignorance. Its only sphere is the Unknowable. It deals with a subject on which, from the nature of our faculties, nothing can be known.

The matter, it is truly said, is one which concerns each and all of us more than any other matter whatever. "It must affect us in all our relations, must determine our conceptions of the Universe, of Life, of Human Nature, and influence our ideas of right and wrong." For if a true reconciliation of Religion and Science "must cause a revolution of thought, fruitful in beneficial consequences," it is no less plain that a false reconciliation, which extinguishes religious faith, and places a blind Fate on the throne of the universe, must be one of the worst of all calamities that can afflict and deceive mankind.

The doctrine in question makes a double affirmation. First, the Being of God, and every other religious dogma, cannot be proved. And next, it cannot even be really believed. Those who think they believe in creation and a Creator are victims of an illusion, which the writer undertakes to explain to them. They are deceived by "a symbolic conception, which cannot be realized in consciousness." Self-existence, the Being of God, an act of creation, are pseud-ideas, and unthinkable.

We may well be curious to learn the nature of this argument, sublime in its audacity, which reduces Christianity, Theism, and every definite religious faith, to ashes in a moment. It consists of two parts, the first original, and the other borrowed from Sir William Hamilton's Philosophy, and Dean Mansel's Lectures on the Limits of Religious Thought.

The original argument is this. Atheism, Pantheism, and Theism are three rival theories, which undertake to explain the great problem of the origin and existence of the universe. But each of them alike is a failure. Atheism fails, because it affirms the self-existence of the universe, and self-existence is unthinkable. "By no mental effort can we form a conception of existence without a beginning."

Pantheism is the doctrine that the universe is self-created. It fails for two reasons. We cannot conceive of potential existence passing into real by some inherent necessity, as self-creation would require us to do; and self-existence, the unthinkable idea, is involved as before for the potential universe.

Theism, again, involves a paralogism, and three unthinkable ideas. The paralogism is the attempt to explain creation by analogy with the work of a human workman. There is no analogy, but contrast, since the workman makes use of pre-existing materials. The three unthinkable ideas are, first, the creation of matter out of nothing; next, that of space; and lastly, the idea of self-existence, as in the two other theories.

Thus the doctrine of the Unknowable, by the method of exclusion, becomes the only possible alternative. In other words, the negative creed, that of the origin of the universe, and the hidden Source of all its changes, nothing whatever can be known.

What shall we say, then, of this fourth alternative, Religious Nihilism? Practically, it is the equivalent of dogmatic Atheism. Logically it differs, and goes one step further in error and self-contradiction.

Theism, as a practical doctrine, affirms that there is a True and Living God, the Creator and Moral Governor of the world, who may be known by every creature endued with reason, and ought to be known, obeyed, loved, and worshipped. Such a God the Nihilist rejects and denies as completely as the dogmatic Atheist. We cannot know one who is unknowable. We cannot obey one, of whom we cannot tell that he has given any commands. We cannot honour, love, or reverence one, of whose nature we have no conception, of whom we do not know whether he is good or evil, or has any moral character whatever.

The doctrine, then, practically, is Atheism under a disguised name. Logically, it is one degree more unreasonable. Theism, Pantheism, and Atheism, all agree that there must be self-existence somewhere. The first ascribes it to a good, holy, perfect Being. The third ascribes it to the countless multitude of material atoms, and supposes these lifeless, self-created or self-existent atoms, to manufacture life and thought by their own combinations. The second also ascribes self-existence to every part of the world, but at the same time ascribes to these a common life and unity, so that all phenomena are modes of one all-pervading soul of the world. Nihilism condemns all three alike for the truth they hold in common, that "there must be self-existence somewhere." This assumption, Mr Spencer says, whether made nakedly or under a disguise, is "equally vicious, equally unthinkable." Yet he admits in the same sentence that the assumption is one "which it is impossible to avoid making." The common fault, then, for which the three rival doctrines are condemned, is that they do what no one can help doing, or believe in "self-existence somewhere."

The peculiar excellence of the doctrine of the Unknowable is that it does what its own author declares no one can do, admits self-existence nowhere. A strange foundation indeed for a new and improved philosophy!

Theists do not, as Mr Spencer alleges, overlook the contrast between creation out of nothing, and the arrangement of existing materials in a Cosmos or well-ordered Universe. It is to the second, not to the first, that they apply the analogy of a human workman, and with perfect reason. They do not deny that the creation of matter out of nothing is mysterious, and has no analogy in the works of men. But mystery is not self-contradiction. The doctrine rests on three simple truths—that being does exist—that hence there must be self-existence somewhere—and that all things around us, and our own minds, have characters opposed to self-existence; that is, weakness, limitation, mutual dependence, and continual change. The idea of creation is definite and intelligible. "By faith we *understand* that the worlds were created by the word of God." But the idea is also a mystery, seen only in part, and therefore we are said to understand it by faith alone.

A further disproof of Theism is drawn from our conception of Space. "If creation were an adequate theory as to matter and mind, it must apply also to space. This must be made as matter is made. But the impossibility of conceiving this is so plain that no one dares assert it."

This assertion, however, is contradicted almost as soon as made. In the opening of the very next chapter the Kantian view of space, as subjective, or a form of human thought, is one of four alternatives, all of which have been or may be held, but are all unthinkable. The

Nihilist, then, must falsify his own creed, and think what he says cannot be thought, before he can derive any argument from the nature of Space to disprove the Christian doctrine of Creation.

The main part of the reasoning, however, in favour of this doctrine of the Unknowable, is borrowed from Sir W. Hamilton's Philosophy, and Dean Mansel's Bampton Lectures. How strange, that the basis of a theory which pronounces all Theology to be made up of pseud-ideas, unthinkable dreams of men self-deceived, should be found in Lectures not only written professedly in defence of Christianity, but welcomed at the time in many quarters with great applause, as a valuable contribution to the cause of Christian faith!

The Lectures in question were no sooner published, and welcomed with high praise by many critics, than protests against their main doctrine and their dangerous tendency were loudly and vigorously made. Professor Maurice was one of the first to offer an indignant protest in his work called 'What is Revelation?' This led to a prompt reply from the Lecturer, and to a rejoinder by Professor Maurice in a second work. The style of the discussion was not faultless on either side, and there seemed some danger of its degenerating into a personal quarrel. But while I differ from much in Professor Maurice's strictures, and in his own counter-statement, and regret the tone of his first work, on the main issue I have no doubt that truth and reason were on his side. The proffered defence of revealed religion was really, however undesignedly, an entire betrayal of its cause, and would make any real defence of it impossible. Dr M'Cosh, of a very different school from Professor Maurice, and an able metaphysician, took substantially the same view.

He remarked on the Hamiltonian doctrine, which forms the entire warp of the Lectures, that "it prepares the way for a Nihilist philosophy, and leaves no ground from which to repel the attacks of religious scepticism."

In two reviews of the Lectures, soon after they appeared, in the 'Christian Observer,' and more recently in a small work, 'The Scripture Doctrine of Creation,' I have taken my share already in the protest which many others have raised. I remarked that the principles, from which the Lecturer deduced his conclusions in favour of Christianity, pointed logically to conclusions entirely different, and highly dangerous. Others, I said, would not fail to trace them to their true issue; and we should then be left stranded on a barren and dreary shore of universal religious Pyrrhonism, still more deadly than those other forms of Rationalism, which the eloquent writer sought to overthrow.

The use which Mr Spencer makes of the Bampton Lectures in the work I am now examining exactly fulfils this anticipation. He quotes largely from them, to establish his own creed of perfect Religious Nihilism. And he does this, he says, for two reasons, that the mode in which the doctrine is presented cannot be improved, and also that since the Lecturer is writing in defence of the current theology, his reasonings will be more acceptable to the majority of readers. He then proceeds to quote four pages in one chapter and three in another, out of sixteen or twenty in the Lectures, and these together form the main substance of the whole argument. I shall quote here a few sentences only, to shew its general nature.

"These three conceptions, the First Cause, the Absolute, the Infinite, all equally indispensable, do they not imply contradiction, when viewed as attributes of one and the same Being? A Cause, as such, cannot be absolute, and the Absolute, as such, cannot be a cause. The cause, as such, exists only in relation to an effect: the cause is the cause of the effect, and the effect is the effect of the cause. On the other hand, the conception of the Absolute implies a possibility of existence out of all relation ! We attempt to escape from this contradiction by introducing the idea of succession in time. The Absolute exists first by itself, and afterward becomes a cause. But here we are checked by a third conception, that of the Infinite. How can the Infinite become that which it was not from the first? If causation is a possible mode of existence, that which exists without causing is not infinite, that which becomes a cause has passed beyond its former limits....The Absolute cannot be conceived as conscious, neither can it be conceived as unconscious. It cannot be conceived as complex, neither can it be conceived as simple. It cannot be conceived by difference, neither can it be conceived by the absence of difference. It cannot be identified with the universe, neither can it be distinguished from it."

"The fundamental conceptions of Rational Theology being thus self-destructive, we may naturally expect to find the same antagonism in their special applications. How can Infinite Power be able to do all things, and Infinite Goodness be unable to do evil? How can Infinite Justice exact the utmost penalty for every sin, and Infinite Mercy pardon the sinner? How can Infinite Wisdom know all things, and Infinite Freedom be at liberty to do or to forbear? How is the existence of evil compatible with that of an Infinitely Perfect Being? For if He wills it, He is not infinitely good; and if He wills it not, his will is thwarted, and his sphere of action limited."

"To sum up the argument, the conception of the Infinite and the Absolute, from whatever side we view it, appears compassed with contradictions. There is contradiction in supposing such an object to exist, either alone or with others, and in supposing it not to exist; in conceiving it as one and as many, as personal and as impersonal, as active and as inactive, as the sum of all existence, and as a part only of that sum."

The deduction of Mr Spencer from these premises is easy and plain. A religious creed, he says, is an *a priori* theory of the universe. Each particular creed asserts two things, that there is something to be explained, and that such and such is the explanation. The latter part is doubly disproved; first, by the discord of these explanations; and next, by the proof in the Lectures that all of them alike are inconceivable, and involve manifold self-contradiction. Atheism, Pantheism, and Theism, are thus alike unthinkable. One element alone survives, which all creeds and religions have in common, that there is a problem to be solved, a something to be explained. The final "soul of truth" which constitutes the reconciliation of Religion and Science, is this doctrine, that an unknowable something exists, but is for ever unknowable, and that "the existence of the world with all it contains and all that surrounds it, is a mystery ever pressing for interpretation." "The analysis of every possible hypothesis proves, not simply that no hypothesis is sufficient, but that none is even thinkable."

But if Mr Spencer finds the best materials for his theory of complete religious Nihilism, where we should least have expected it, in Bampton Lectures designed expressly for the defence of Christian Faith, the leading work of another non-Christian philosopher, of equal reputation with his own, if not still greater, supplies us with an antidote and refutation. There is thus a strange confusion and crossing of parts in this intellectual and moral controversy. Mr Mill, in his Examination of Sir W. Hamilton's Philosophy, takes up those very statements of Sir William himself, and of the Bampton Lecturer, on which Mr Spencer mainly relies, and gives them, I think, a clear, plain, and decisive refutation.

The first maxim implied in the doctrine of the Unknowable is that inconceivableness or unthinkableness is a term of one meaning only, and the same with self-contradiction. On this assumption Mr Mill, replying to Hamilton, writes as follows:

"Our author goes on to repeat his argument used in his reply to Cousin, that infinite space is inconceivable, because all the conception we are able to form of it is negative, and a negative conception is the same as no conception. The Infinite is conceived, only by thinking away every character by which the finite is conceived. To this I oppose my former reply. Instead of thinking away every character of the finite, we think away only the 'idea of an end or boundary. The proposition is true of the Infinite, as a meaningless abstraction, but it is not true of infinite space. In trying to form a conception of that, we do not think away its positive characters. We leave to it all that belongs to it as space, its three dimensions, with their geometrical properties. We leave to it a character which belongs to it as infinite, of being greater than any finite space. If an object which has these well-marked positive attributes is unthinkable, because it has a negative attribute as well, the number of thinkable objects must be remarkably small."

"In other passages," Mr Mill continues, "Sir W. H. argues that we cannot conceive infinite space, because we should require infinite time to do it in." This is precisely Mr Spencer's argument on a kindred subject, the infinite divisibility of matter or space. To go through the process would need infinite time (p. 50). Mr Mill's reply is clear and decisive.

"It would of course require infinite time to carry our thoughts in succession over every part of infinite space. But on how many of our finite conceptions do we think it necessary to perform such an operation? Let us try the doctrine on a complex whole, short of infinite, such as the number 695,788. Sir W. H. would not, I suppose, maintain that this number is inconceivable. How long does he think it would take to go over every separate unit of this whole, so as to obtain a perfect knowledge of that exact sum, as different from all others, greater or less? Would he say that we can have no conception of the sum, till this process is gone through? We could not indeed have an adequate conception. Accordingly we never have an adequate conception of any real thing. But we have a *real* conception, if we conceive it by any of its attributes, which are sufficient to distinguish it from all other things. ...If then we can obtain a real conception of a finite whole, without going through all its component parts, why deny us the conception of an infinite whole, because to go through them all is impossible? Not to mention that even in the case of the finite number, though the units are limited, the possible modes of deriving any number from other numbers are numerically infinite. And as these are necessary parts of an adequate conception of it, to render our conception even of this finite whole perfectly adequate would also require infinite time."

Such is Mr Mill's conclusive reply to the Pyrrhonist as to Space or Number. The same answer plainly applies in Theology. If Religion is nescience, because it deals with an infinite object, and such an object cannot be adequately conceived or known within a finite time, then Physics are nescience for the very same reason. There is no object, though finite, of which all the relations, either within itself, or to other objects, can be exhaustively known by any finite mind. The number two is one of the simplest objects of thought. But to know perfectly either its square root or its common logarithm in their ratio to unity, since the number of decimals in either is infinite, must lie beyond the reach of any finite understanding.

Again, to the reasoning of Sir W. Hamilton and Dean Mansel, adopted by Mr Spencer, that an Infinite Being must be wholly unknowable, Mr Mill replies as follows:

"But is a conception, by the fact of its being a conception of something infinite, reduced to a negation? This is quite true of the senseless abstraction, 'the Infinite.' That is purely negative, being formed by excluding from the concrete conceptions, classed under it, all their positive elements. But in the place of 'the infinite' put 'something infinite,' and the argument collapses at once. 'Something infinite' is a conception which, like most of our complex ideas, contains a negative element, but positive elements also. Infinite space is an example, and infinite duration. It would surprise most persons to be told that eternal life is a purely negative conception, that immortality is inconceivable. Those who hope for it have a very positive conception of what they hope for. Between a conception which, though inadequate, is real as far as it goes, and the impossibility of any conception, there is a wide difference."

In the "Scripture Doctrine of Creation," the same view substantially is maintained, though in a slightly different way. "The Absolute and the Infinite," I have remarked, "are not the same with absoluteness and infinity. Each name is twofold. The article holds the place of a substantive to the epithet that follows. It implies some object of thought, differenced by that epithet from all other such objects. The Absolute, then, by the force of the term, is first of all some definite object of thought; and next, one defined by freedom from various imperfections of other objects. The name implies a relation to human intelligence. To make the absoluteness exclude any such relation turns the name into a chimera and self-contradiction. The Infinite, in like manner, is first a definite object of thought; and next, one which is free from the limits our experience assigns to other objects. To define it, then, as a mere negation of limits, with no affirmation of real and unlimited Being, contradicts the phrase, and robs it of all meaning....The Absolute and the Infinite, again, are not conceptions purely negative. Rather, of all conceptions they are the most positive and real. For Being is a positive idea, but a limit beyond which there is the absence of being, is negative. Thus Being is a positive idea, which the Absolute, the Infinite, shares with the finite or conditioned. But there is then the added element of fulness and perfection of being, such as excludes not-being, limitation, and imperfection alone."

The agreement here in substance with Mr Mill's argument is complete. In the detail where we differ I think that he has committed an oversight, that his remarks apply to absoluteness and infinity, or to the phrases if taken plurally, but that the Absolute, the Infinite, do naturally denote one true, All-perfect Being.

Sir W. Hamilton and Dean Mansel maintain, in opposition to Cousin and many German writers, that a Rational Theology is impossible, because God is infinite and absolute, and that these are ideas contradictory of each other, and each involving also a self-contradiction. The Lectures strive to escape from the logical results of this doctrine, by stating that the Bible reveals, not what God is, but how He wills that men should think of Him. But this view is fatally opposed to the theory which it is meant to reconcile with Christianity. A Being cannot be wholly unknown, of whom we know that He has a will, that He is distinct from his creatures, that He has moral preferences among the opinions of men, and requires them to think of Him in one particular way. And the view also casts a dark cloud on the divine character. It ascribes to the Supremely Good and Wise the desire that his creatures should accept a mere shadow for a reality, because He is unable to give them any genuine revelation.

Mr Spencer carries out the doctrine to its true and proper issue. He refuses to allow that it can be our duty to think of God in one way rather than another, as a person, or good and holy, which would imply, on his theory, "an eternal war between our intellectual faculties and our moral obligations." If the doctrine were sound, our only duty would be absolute neutrality, and neither to affirm nor deny anything in a region of thought, where nothing can be known.

Eternal life can never consist in a knowledge which is only a fiction and a shadow. This grand objection the Philosophy of the Conditioned cannot overcome without renouncing all Christian faith, or else committing suicide in some other way. For if the Absolute, as that philosophy affirms, has a necessary existence, and is therefore a name of the true and eternal God, and still can neither be known nor conceived at all, the sceptical conclusion must be irrefragably sure. Theism sinks to the level of an unproveable hypothesis, and God, if He exists, must be viewed as a Being of whom nothing whatever can be known.

Another main element of this Doctrine of the Unknowable, which it shares with Sir W. Hamilton's Philosophy and the Bampton Lectures, is the denial of man's capacity to know anything of the moral attributes of God; or if we vary the name for the sake of reverence, of the Infinite or the Absolute, styled otherwise "the utterly inscrutable Power which the universe manifests to us." Thus we read in the Lectures:

"If we know not the Absolute or the Infinite at all, we cannot say how far it is or is not capable of likeness or unlikeness to the Relative and Finite. We must remain content with the belief that we have that knowledge of God, which is best adapted to our wants and training. How far that knowledge represents God as He is, we know not, and we have no need to know."

Mr Spencer, with far greater consistency, draws the following inference, and discards, as a plain contradiction, a theology of regulative truths, in a field of thought where nothing can be known:—

"After it has been shewn how, by the very constitution of our minds, we are debarred from thinking of the Absolute, it is still asserted that we ought to think of the Absolute thus and thus. In all imaginable ways the truth is thrust upon us that we are not even permitted to conceive the Reality which is behind the veil of appearances.

And yet it is said to be our duty to believe, and so far to conceive, that this Reality exists in a certain defined manner! Shall we call this reverence, or call it the reverse? Volumes might be written on the impiety of the pious."

Such is the grave rebuke, in Mr Spencer's First Principles, of the Bampton Lecturer's attempt to escape from the logical result of a theory held in common by both, and to patch a regulative Christianity on the seamless robe of the Philosophy of the Unconditioned, or the Doctrine of the Unknowable. And now what sentence, in turn, does Mr Mill pronounce upon both? "My opinion," he says, "of this doctrine (namely, that nothing can be known or understood of moral attributes in a Supreme Being) in whatever way presented, is that it is simply the most morally pernicious doctrine now current, and that the question it involves is, beyond all others which now engage speculative minds, the decisive one between good and evil for the Christian world. I think it supremely important to examine whether the doctrine is really the verdict of a sound metaphysic. I think that the conclusion not only does not follow from a true theory of the human faculties, but is not even correctly drawn from the premises from which it has been inferred." (Ex. p. 113.)

Mr Mill then quotes one of the main passages in the Lectures, which Mr Spencer has chosen for the basis of his whole system, and reasons on it as follows:

"The whole argument for the inconceivability of the Absolute or the Infinite is one long *ignoratio elenchi*. It does not prove that we cannot know an object absolute or infinite in some particular attributes, but only that we cannot know an abstraction, called the Absolute or the Infinite, which is supposed to have all attributes at once. He expressly identifies it with Hegel's absolute being, which contains in itself 'all that is actual, even evil included.' That which is conceived as absolute and infinite, says Mr Mansel, must be conceived as containing within itself the sum, not only of all actual, but all possible modes of being. One may well agree with him that this farrago of contradictory attributes cannot be conceived. But what shall we say of his equally positive averment, that it must be believed ? If this be what the Absolute is, what does he mean by saying that we must believe God to be the Absolute ?"

Again, after quoting from the Lectures the passage (p. 71) on the Infinite, exactly parallel to the one (pp. 58, 59) which Mr Spencer has taken for the chief foundation of his Doctrine of the Unknowable, Mr Mill criticises it in these words:

"Here certainly is an infinite whose infinity does not seem to be of much use to it. But can a writer be serious, who bids us conjure up a conception of something which possesses infinitely all conflicting attributes, and because we cannot do this without contradiction, would have us believe that there is contradiction in the idea of Infinite Goodness or Infinite Wisdom ? Instead of 'the Infinite' substitute 'an Infinitely Good Being,' and the argument reads thus :—If there is anything which an infinitely good Being cannot become, if He cannot become bad, there is a limitation, and the goodness cannot be infinite. If there is anything which He is, namely, good, He is excluded from being any other thing, as from being wise or powerful....He says, we are compelled by the constitution of our minds to believe in the existence of an Absolute and Infinite Being. Such being the case, I ask, Is the Being whom we must believe to be absolute and infinite, infinite and absolute in the sense these words bear in his definition of them ? If not, he is bound to tell us in what other meaning. He has either proved nothing, or vastly more than he intended. The contradictions he asserts to be involved in the notion do not follow from an imperfect mode of apprehending them, but lie in the definitions. If therefore he would escape from the conclusion that an Absolute and Infinite Being is impossible, it must be by affirming, with Hegel, that the law of contradiction does not apply to the Absolute, and that with respect to it contradictory propositions may both be true."

Such, then, according to Mr Mill, is the intellectual or logical nature of the reasoning which Mr Spencer adopts as the grand foundation of his philosophical system. It confounds the conception of a Being infinitely good and wise with that of one which possesses all incongruous attributes in an infinite degree, and with an Absolute which includes all things evil; and then, because these chimeras are inconceivable, infers that nothing can be known of the Infinitely Good and Wise.

On the moral aspect of the same theory, which denies that we have any knowledge how far what we call moral characters can apply to an Infinite Being, Mr Mill remarks as follows :

"Here, then, I take my stand on the acknowledged principles of logic and morality, that when we mean different things, we have no right to call them by the same name. Language has no meaning for the words, just, merciful, benevolent, save that in which we predicate them of our fellow creatures, and unless that is what we intend to express by them, we have no business to employ the words. If in affirming them of God we do not mean to affirm these very qualities, only greater in degree, we are neither philosophically nor morally entitled to affirm them at all. I grant that we cannot adequately conceive them in one of their elements, their infinity. But we can conceive of them in their other elements. Anything carried to the infinite must have all the properties of the same thing, as finite, except the finiteness. What belongs to infinite goodness as infinite or absolute I do not pretend to know. But I know that infinite goodness must be goodness, and that what is not consistent with goodness is not consistent with infinite goodness. If, in ascribing goodness to God, I do not mean the goodness of which I have some knowledge, but an incomprehensible attribute of an incomprehensible substance, which for aught I know may be wholly different from that which I love and venerate, what do I mean by calling it goodness, or what reason have I for venerating it? Unless I believe God to possess the same moral attributes, which I find, in however inferior a degree, in good men, what ground of assurance have I of God's veracity? All trust in a Revelation presupposes a conviction that God's attributes are the same, except in degree, with the best human attributes. If I call any Being good or wise, not meaning the only qualities which the words import, I am speaking insincerely. I am flattering him by epithets which I fancy that he likes to hear, in the hope of winning him over to my own objects."

"The proposition, that we cannot conceive the moral attributes of God in such a way as to be able to affirm of any assertion that it is inconsistent with them, has no foundation in the laws of the human mind. If admitted, it would not prove that we should ascribe to God attributes bearing the same name as human qualities, only not to be understood in the same sense. It would prove that we ought not to ascribe any moral attributes to God at all, inasmuch as no moral attributes known or conceivable by us are true of Him, and we are condemned to absolute ignorance of Him as a moral being."

There are two or three sentences in the passage from which I have made these extracts, which I regret deeply, and from which I wholly disagree. But the reasonings I have now quoted I believe to be just and sound both in logic and morality. And if true, they overturn Mr Spencer's Doctrine of the Unknowable from its very foundations. They prove that he has borrowed from Sir W. Hamilton and Dean Mansel, and carried out to their logical results, principles which involve the highest degree of self-contradiction, and confound two impossible chimeras, inclusive of all kinds of evil and folly, with the true and Christian conception of an Infinitely Good and Absolutely Perfect Being. And the proof goes further. It shews clearly that the doctrine thus advanced, if it be not, as he affirms, "the most morally pernicious doctrine now current," at least may fairly compete with two or three other rivals for that bad preeminence. For it plainly shuts up all mankind in total ignorance and darkness on all religious matters, and forbids them to have any faith in a Creator or Moral Governor of the world. It denies them the right to ascribe to "that Power which the universe manifests to us," and which it proclaims to be "utterly inscrutable," any kind of goodness or moral perfection, or any character which could have any claim on our love, worship, or obedience. It is thus a doctrine of despair, under which absolute moral and religious darkness is made to settle down upon the whole universe, with no possible gleam of light, for ever and ever.

But now let us inquire how far this Doctrine of the

Unknowable, the proposed treaty of peace between Religion and Science, by which we are to escape from the alleged contradictions of Christianity, Theism, and every kind of positive religious doctrine, is itself consistent or conceivable. It is embodied in one short sentence. "The widest, deepest, and most certain of all facts, is that the Power which the universe manifests to us is wholly inscrutable."

Such briefly is the sum of the whole doctrine, and it contains five or six self-contradictions. And first, do we know that this Power exists? So we are afterwards assured. We are told that it is an indestructible belief, that "it cannot cease till consciousness ceases, and has the highest validity of any." If so, we know one thing with regard to this Unknowable, that it has a real existence. Do we know that it is not a mere attribute of something else? This is a second degree of knowledge. Do we know that it is One Power, and not a mere medley of many independent persons or things? This will be a third degree of knowledge. Do we know that it is rightly described as a Power at all, and is not rather weak, impotent and powerless? This will be a fourth element. Does the universe manifest it to us? Then clearly it cannot be wholly unknown. Entire hiddenness is contradicted alike by either partial or total manifestation. Is this Power distinct from the universe which manifests it to us, or is it another name for the universe itself? If distinct from it, as the axiom implies, this will be a sixth element in our knowledge of this Unknowable Something. That it exists, that it is not an attribute, but either thing or person, that it is one person or thing, and not many, that it is distinct from the universe which manifests it, and that it is really manifested by the universe, that it is a Power and not a mere Impotence, are six truths affirmed concerning it in the very definition, which speaks of it as utterly inscrutable and unknown. And if we add to these the statements which presently follow, that it stands in a relation of contrast to the Relative (p. 91), that it is "the persistent body of a thought to which we can give no shape, and the object of an irresistible belief" (p. 93), that it is "a something, the concept of which is formed by combining many concepts, deprived of their limits and conditions" (p. 95), that it is "an actuality lying behind appearances" (p. 97), that it is in such close relation to the relative realities, that every change in one may be viewed as representing an answering change in the other, so that the relatives and absolutes are practically equivalent (p. 162), and finally, that more or less constant relations in the absolute beyond consciousness are matter of experience, and generate like relations in our states of consciousness (Test of Truth, p. 548), we may see the force of Mr Mill's satirical remark, that the doctrine recognizes as attainable a surprising and almost prodigious amount of knowledge of the Unknowable.

Once more, the residuary truth, which alone remains, when all religious creeds have been swept away, and on which there is said to be entire unanimity amongst devotees of every name, and sceptics of every school of philosophy, is thus expressed—that the existence of the world, with all that it contains, and all that surrounds it, is a mystery ever pressing for interpretation.

Now it is quite natural and consistent, in all those who believe that there is a God, the Creator and Moral Governor of the world, and that some knowledge of his nature, and his relations to his own creatures, is attainable and highly desirable, to seek earnestly to attain it; and if any measure of it has been attained, to desire its increase, and strive upward into fuller light. It is no less natural that they should reflect their own feelings and desires on the object of their pursuit; so that the mystery, into which they long for clearer insight, may be said, by an easy figure, to be for ever pressing for interpretation. But in accepting the creed of religious Nihilism, along with Christian faith and every other religious doctrine, this residuum itself must vanish and disappear. How can a mystery press for interpretation, of which we know and are sure that no interpretation will ever be found? Once adopt this negative creed of Nihilism, this doctrine of the unknowable, and the pressure must wholly cease. Labour, effort, and earnest striving are utter folly, when there is no hope of the least success. All religious creeds, so far as they retain the least element of truth, may be unanimous in their common sense of the deep mystery of God and the universe, and in cherishing a longing wish to see it more clearly, and gain a deeper insight into its meaning. Modern Nihilism is the solitary exception to this unanimous instinct of all truly thoughtful minds. To be consistent with itself it can "admit no such radical vice in the constitution of things as an eternal war between our intellectual faculties and our moral obligations." Since it is impossible for us ever to know or learn anything of this great mystery of the universe, our only religious duty is to conform our feelings to the actual and fated constitution of things, and to rest content and satisfied with total darkness. Under the new philosophy the cry of the patriarch long ago—" O that I knew where I might find

Him! that I might come even to His seat"—was an utterance of pure and unmingled folly. For with regard to the faintest glimmer of insight into this great mystery, the nature of God, and his relations to his intelligent creatures, this motto is inscribed over its gloomy portal—

"All hope abandon, ye who enter here."

CHAPTER II.

ON ULTIMATE IDEAS IN PHYSICS.

THE words of the Apostle, "we know in part," are the keynote of all true Philosophy. They apply equally to the lowest and the highest field of human thought, to the study of number and lifeless matter, and to the knowledge of God, the Supreme Creator. We know. We are condemned nowhere to utter nescience and total darkness. But we know in part. Our knowledge is everywhere surrounded by mystery, and much remains unknown. In every field of thought the assertion that we can know nothing at all is a degrading superstition; while a claim to perfect knowledge, excluding mystery, is the extreme of presumptuous folly.

The Philosophy of the Conditioned, in Sir W. Hamilton, and its equivalent, the Doctrine of the Unknowable, in Mr Spencer's Philosophy, consists in a direct reversal of this fundamental truth. The first maintains that we must believe the unconditioned to exist, but that, from the very laws of thought, we can know nothing of it whatever. The latter adopts the same view, and carries it out to its logical conclusion. The contrast between Theology and Physics, which alone is honoured with the name of Science, is affirmed to be the same with the contrast of the unknowable and the knowable. All recognize that the nature of God is deeply mysterious. All, or nearly all, believe that great and real advances have been made in physical science, and the knowledge of outward things. A true philosophy will complete these admitted facts by the double doctrine that in Theology much may and ought to be known, and that in Physics much is, and will remain to the last, mysterious. The new philosophy, to be consistent, must hold the two contradictory falsehoods, that in religion all is mystery, with no possibility of real knowledge, and that in Physics we may escape from mystery altogether, and attain to pure and perfect knowledge, in which nothing remains unknown.

The chapter on Ultimate Religious Ideas is designed to prove the first of these doctrines. I have now shewn, and confirmed it by Mr Mill's carefully reasoned remarks, that the whole is one continuous fallacy. Its closing paragraph is as follows:

"Here, then, is an ultimate religious truth of the highest possible certainty, a truth, in which religions in general are at one with each other, and with a philosophy antagonistic to their special dogmas. This truth, respecting which there is a latent agreement among all mankind, from the fetish-worshipper to the most stoical critic of human creeds, must be the one which we seek. If Religion and Science are to be reconciled, the basis of reconciliation must be this deepest, widest, and most certain of all truths, that the Power which the universe manifests to us is utterly inscrutable."

Surely a more prodigious statement was never made, much less taken for the basis of a new philosophy. The one dogma in which all mankind agree, however they differ in details, that no religious dogma whatever is either thinkable or credible! The one way of reconciling Religion and Science, the doctrine that they have nothing in common, and are perfect opposites—Science being the knowledge of all things knowable, that is, physical changes, and Religion merely another name for our helpless and hopeless ignorance of the Unknowable!

This grand discovery, however, though styled "the widest, deepest, and most certain of all truths," is contradicted in the very words chosen to describe it. "The Power which the universe manifests to us is utterly inscrutable." But of this unknowable we cannot know whether it is Power or Impotence, one power or impotence, or many powers or impotences, whether it be thing or person, or things or persons. Of one thing only we may be sure, that the universe cannot manifest it to us, because it is "utterly inscrutable," and must remain for ever unmanifested and unknown.

To complete the theory, the attempted proof that Religion is mystery without any knowledge should be followed by a like proof that Physics is knowledge free from all mystery. Strange to say, the next step in its actual development is just the reverse. Chapter the Third, on Ultimate Scientific Ideas, is occupied with a proof that Space, Motion, Time, Matter, Force, the main ideas in Physics, are unthinkable and full of contradiction, like Self-existence, the First Cause, the Absolute and the Infinite, the primary conceptions of Theology. Thus the contrast between Theology as pure mystery, and Physics, as free from mystery, and pure science, expires even before it is fully born. In each it is allowed, and even argued, there is the same presence of mysteries, which we cannot thoroughly expound, and which it is striven to convert into direct self-contradiction. Now if these do not exclude, but accompany, real and progressive knowledge in one case, how can they possibly exclude it in the other? It is owned that real knowledge is attainable, and has been attained in growing measure, in Physical Science. The inference is plain. A real, and even a growing knowledge, in spite of its mysteriousness, must be no less attainable in Sacred Theology, that highest field for the exercise of human thought, the knowledge of the Holy One, which is true wisdom.

The sixth chapter of Mr Mill's Examination of Sir W. Hamilton's Philosophy is occupied with this theory of the Antinomies, which Mr Spencer has borrowed, and taken for the basis of his own work. He shews, I think, very clearly, that the doctrine rests on an ambiguous and deceptive use of the words, inconceivable and unthinkable, which admit of two or three different meanings. Of propositions and their direct negatives one must be true, the other false. If we call them both unthinkable, we must use the word in two widely different senses. In one sense it denotes simply mystery, a truth known in part, and too large and vast to be fully understood. In the other it denotes, or may denote, a self-contradiction, a proposition of which the elements are inconsistent, and exclude each other. In the first sense, of course, all primary or ultimate truths are unthinkable or incomprehensible. But Mr Mill justly remarks—"If all general truths which we are most certain of are to be called inconceivable, the word no longer serves any purpose. A truth which is not inconceivable in either of the received meanings of the term, a truth which is completely apprehended, and without difficulty believed, I cannot consent to call inconceivable, because we cannot account for it, or deduce it from a higher truth."

Mr Spencer's reasoning is of this kind. Space and Time cannot be thought of objectively, either as entities or attributes of entities. We cannot assert of them either limitation or absence of limitation. We cannot form any mental image of unbounded space, nor imagine bounds beyond which there is no space. We cannot conceive either its limited or unlimited divisibility. So also of Time. To call them subjective, that is, laws or conditions of the conscious mind, is to escape from great difficulties by rushing into greater. The proposition that they are purely subjective "cannot by any effort be rendered into thought, but stands merely for a pseud-idea." If they *are* forms of thought, they can never be thought of, since it is impossible for any thing to be at once the form and matter of thought. They are therefore wholly incomprehensible. Our knowledge of them is total ignorance.

So also of Matter. We cannot think of it consistently, either as finitely or infinitely divisible, as completely solid, or as composed of discrete solid atoms never in contact, or as unextended centres of force. "Frame what suppositions we may, we find nothing but a choice between opposite absurdities."

Our ideas of Motion also are illusive. Absolute Motion cannot even be imagined, much less known. Motion taking place apart from those limitations of place, with which we usually associate it, is totally unthinkable. "While we are obliged to think that there is an absolute Motion, we find absolute Motion incomprehensible." A like difficulty attends the transfer of Motion. "A striking body has not transferred a *thing*, and it has not transferred an *attribute*. What then has it transferred?" The transition from Motion to Rest is inconceivable. The smallest movement is separated by an impassable gulf from no movement. The least conceivable motion is infinite as compared with absolute rest. All efforts to understand Motion bring us thus to alternate impossibilities.

The same is said to be true of Force. It is absurd to think of it as like our sensations, and yet necessary so to think of it, to realize it at all. How can we understand the connection between Force and Matter? The hypothesis of Newton, as well as of Boscovich, supposes one thing to act on another through absolutely empty space, a supposition which cannot be represented in thought. By introducing a hypothetical fluid the difficulty is merely shifted, and transferred to the constitution of that fluid. "We are obliged to conclude that Matter, whether ponderable or imponderable, acts on matter through absolutely vacant space, and yet this conclusion is absolutely unthinkable."

Another antinomy is found in the law of attraction and repulsion as the inverse square. "We are obliged to say that the antagonist forces do not both vary inversely as the squares of the distance, which is unthinkable; or else that Matter does not possess the attribute of resistance, which is absurd."

As to Mind, and states of consciousness, it is just the same. We cannot say that the series is infinite, for infinity is inconceivable. Nor yet finite, for we have no knowledge of either end. We can neither conceive them nor perceive them. To represent the termination of consciousness as occurring in ourselves is to think of ourselves as contemplating the cessation of the state of consciousness, and this implies the continuance of consciousness after its last state, which is absurd. Nor is the subject of consciousness thinkable. "The fundamental condition of all thought is the antithesis of subject and object. Now if the object perceived is self, what is the subject that perceives? If it is the true self that thinks, what other self can it be that is thought of? Cognition of self implies a state in which subject and object are identified, and this Mr Mansel rightly holds to be the annihilation of both. Thus the personality, of which each is conscious, is a thing which cannot be known at all. The knowledge of it is forbidden by the very nature of thought."

Here is indeed a wide and large basis for the theory of Religious Nihilism and Physical Fatalism. The building does not rest on an elephant or a tortoise, like the earth in the Hindu creed, but on a series of self-contradictions, a conglomerate of infinite darkness and confusion. How any consistent scheme of thought can be raised on such a foundation seems of all inconceivabilities the most inconceivable. Of the same doctrine in substance, as taught by Sir W. Hamilton, Mr Mill remarks as follows:

"In the case of each of the antinomies which the author presents, he undertakes to establish two things, that neither of the rival hypotheses can be conceived as possible, and that nevertheless we are certain that one or other of them must be true. I have shewn strong reasons for dissenting from this assertion; and those which our author assigns in its support seem to me quite untenable."

Again—"If the doctrine hold, we cannot predicate any thing of a subject, which we regard as being in any of its attributes infinite. We are unable, without falling into a contradiction, to assert anything, not only of God, but of Time and Space. Considered as a *reductio ad absurdum*, this is sufficient. If because the comprehension of a general notion is finite, any thing infinite cannot without contradiction be thought under it, then a Being possessing in an infinite degree a given attribute, cannot be thought of under that very attribute. Infinite Goodness cannot be thought of as goodness, because that would be to think of it as finite. There must be surely some great confusion of ideas in the premises, where such is the conclusion."

He then concludes his argument. "There would be no difficulty in applying a similar line of reasoning to the case of Time, or any other of the antinomies. In no case mentioned do I believe that he could substantiate his assertion that the conditioned, that is, every object of human knowledge, lies between two hypotheses, both of them inconceivable......The proposition that the Conditioned lies between two hypotheses concerning the Unconditioned, neither of which we can conceive as possible, must be placed in that numerous class of metaphysical doctrines, which have a magnificent sound, but are empty of the smallest substance."

Since, however, three authors of such high reputation as Mr Spencer, Sir W. Hamilton, and Dean Mansel, all take here the same ground, and the last has vehemently and almost contemptuously denied the soundness of Mr Mill's counter-argument, it may be well to sift the question a little further; especially since Mr Spencer, while borrowing the main thoughts from his predecessors, has made some original, and, I think, strangely and grotesquely erroneous additions of his own.

The direct and proper result of the doctrine, common to all three writers, is universal Pyrrhonism. Of the

Absolute or the Infinite nothing whatever can be known. But the proper idea of the Absolute, the Bampton Lectures affirm with Hegel, is all-inclusive, "evil not excepted." Therefore of all and every Being, which must be either the Absolute or a part of it, nothing at all can possibly be known.

Dean Mansel attempts to escape from the consequence of the doctrine, in Theology, by recognizing regulative in the place of speculative truth; that is, resemblances or analogies to the Unknowable; lessons how God wills that we should think concerning Him, though the truth of the Divine Being and Nature must be beyond our reach. But the effort, though well meaning, is vain and futile. On his own principles, there may perhaps be regulations, but not regulative truths, and the regulations must be human, or perhaps superhuman, but not Divine. Of a Being wholly unknowable we cannot possibly know that He has given us a revelation, or wills us to think about Him in one way rather than another. Mr Spencer, who accepts the premises, more logically rejects this compromise of a regulative Theology, and even sets it down as a new example of that impiety and vain presumption of the pious, on which volumes might be written.

But when Theology has been flung into this gulf of utter darkness, how are Physics to be rescued? Under the shadow of this theory, in which God, Space, Time, Matter, Motion, Force, are all unthinkable without a choice between two equal absurdities, how can there be possible knowledge of any kind? Here it is Mr Spencer's turn to be illogical, to tread backward on his own footsteps, and to build up with toil and labour what he has pulled down and laid in ruins. After a supposed proof that Space, Time, Matter, Motion, Force, Consciousness are just as unthinkable and full of contradiction as the bases of Theology, and a regulative Theology has been denounced with bitter scorn, we are introduced forthwith to a Regulative Physics, with a long series of *à priori* and *à posteriori* truths. By this means a new system of the universe is built on the old site, when Christianity and every other form of religious faith have been cleared ignominiously out of the way.

Space, Time, Matter, Motion, and Force are the common subject of Bk. I. ch. III. on the Unknowable, and Bk. II. ch. III. on the Knowable. In the first we are taught that each possible alternative is a pseud-idea, and unthinkable. In the second they experience a philosophical resurrection. We are told the precise contents of each idea, the definite relation in which the relative stands to the absolute, the knowable to the unknowable. Matter is the synthesis of extension and resistance. The first element is derivative, the other primary. Motion at first is an infant conception, which becomes adult and mature, evolved from varied impressions of muscular tension and objective resistance. Force is a conditioned effect of the unconditioned cause. It is a relative Reality, indicating an absolute Reality by which it is produced. We may see clearly, it is said, "the transfigured realism to which sceptical criticism finally brings us round." In short, Space, Time, Matter, Motion, Force are all inconceivable and unthinkable, and every idea we can form of them a pseud-idea. But we can still think of them to such good effect, as to build up from those thoughts a true philosophy, a clear solution of the great problem of the world's unceasing changes. The later dogmatism may give us, then, some help in proving the entire fallacy of those Pyrrhonistic reasonings, which are the basis and outset of the whole work.

I start from this simple premise. If two contradictories are pronounced to be alike inconceivable, the meaning of the title, as applied to both, is not the same. A false alternative may be unthinkable, either as disproved by facts, and therefore incredible as fact, or a self-contradiction, where the elements, verbally united, exclude each other, like a circular square, or two-sided polygon. But an alternative which is true can be inconceivable only in a very different sense. While we may think of it, believe it to be true, and partly apprehend it, we may be unable to comprehend it, or take in the full compass of its true meaning. If alternatives are not contradictory, both may be false, and leave room for a third alternative, which is true, and free from all contradiction, though a full and perfect comprehension of it may exceed the limits of any finite intelligence.

Let us now review the chief data of Mr Spencer's argument.

First, Space and Time are said to be unthinkable, either as subjective or objective. But is the inconceivableness on each side the same? Mr Spencer himself gives a negative answer. "Our belief in their objective reality is insurmountable." "To posit the alternative belief is to multiply irrationalities." The same contrast is unfolded in the Psychology with much force of reasoning. That Space and Time, then, are mere subjective affections of the mind, or forms of thought, may be unthinkable, because it is false. But then the other alternative, that they are objective, must be true, and conceivable in the most proper sense of the term.

But if they are objective, we are told that we must

adopt one of three alternatives; that they are non-entities, entities, or attributes of entities. Reasons against each alternative are then given. But is the list complete and exhaustive? We turn to the later chapter, and find there a fourth alternative, which Mr Spencer adopts and reasons upon as the true and proper view. They are neither entities, nor attributes of entities, but relations between them. "We think in relations. The two main classes of relation are sequence and coexistence. The abstract of all sequences is Time. The abstract of all coexistences is Space. From blank forms of coexistence, from which the coexistent objects are absent, and a building up of these,...results that abstract of all relations of coexistence which we call Space." If so, the alleged proof of its entire inconceivability falls to the ground. The unthinkableness results from ingenious management alone. Three false alternatives are examined; but a true alternative, which the author presently accepts, and on which he builds a process of ingenious and able reasoning, is passed by in total silence.

A third inconceivability follows. We can think of Space and Time neither as limited nor unlimited. Yet surely these are contradictories, and one or other must be true. But here one great ambiguity vitiates the whole reasoning. If Space and Time denote relations of coexistence or of sequence, these must mean either of actual or of possible things. Here, then, we have a key to the difficulty. The created universe, we may assume, is finite both in number, extension, and past duration. Thus Actual Space and Actual Time, the time that has really elapsed, will both be finite. But the possible relations are infinite. However great, being finite, we must conceive it possible for them to be, or to become, greater. And thus Space and Time, when viewed as the sum of all possible relations of coexistence and vastness, or succession, between all possible creatures, are and must be infinite.

The picture in Milton thus admits of a philosophical interpretation. The whole finite creation hangs suspended from the Infinite Reality, whence it derives its being, and is surrounded by an immense void of the infinite possibilities of finite and created existence, still unborn.

Next, matter is pronounced inconceivable on each of three alternatives, a plenum, distinct solid atoms, and centres of force. A plenum I believe, as Lucretius argued long ago, to be inconsistent with the motion of bodies, and makes it impossible. The view of finite atoms, though by no means unthinkable, and held by a multitude of philosophers, from Democritus to Newton, and able authors of our own day, I do not care to defend, as I think it erroneous, and to agree neither with the deepest mental analysis of matter, nor with the course of experimental science. But the unthinkableness of the third alternative I wholly deny. The notion of atoms as unextended centres of force may be mysterious, as are all the deepest truths, but involves no contradiction. I should rather be disposed to maintain that no other view can be thought of, and reasoned out, without proving itself really unthinkable.

The assertion that such centres of force are inconceivable admits a double disproof. First, as a matter of simple fact, they have been conceived, and made the starting-point of strict and consistent reasoning, in many works. One third of Newton's Principia is made up of reasonings based on this very conception. None of the cardinal propositions of the first Book are true or intelligible, unless we refer them to points, and not to extended masses. The law of gravitation, which Mr Spencer strangely asserts to be an *à priori* truth, a necessary result of the nature of space, was only discovered by means of strict and varied reasonings, based on this very hypothesis, of which he ventures to say that to make it or think it at all is "utterly beyond human power."

I believe that I may go further, and say that the inclusion of extension within the units exercising force, instead of referring it to the relations between them, is what is really unthinkable. Forces that vary by any function or power of the distance cannot possibly be referred to anything but points alone. The distance of a point from a plane is really its distance from the point where the vertical intersects the plane. A mass or bulk has not one distance from another mass or bulk. The number of distances is as great as the acting points of the one multiplied by the acting points of the other. Or again, if distance is a relation of coexistence between different units, then the relation cannot belong to the unit itself. Extension implies and requires manifoldness or plurality. The assertion, then, that we cannot conceive centres of force at all, is wholly untrue. The conception is most definite. It has been made the basis of a whole body of dynamical reasonings. All the greater discoveries of physics have grown out of mathematical reasonings, grounded on this conception, and which, without first adopting it at least provisionally, could never have been attained. The conservation of *vis viva*, the only nucleus of truth in the doctrine which Mr Spencer styles the Persistence of Force, is only a hypothetical consequence of the same conception. It is a strange logical error to make the murder of the parent thought a direct preparation for an apotheosis of the child.

Motion is next said to involve two contradictions. First, we cannot conceive it to be either merely relative, or non-relative and absolute. "While we are obliged to think that there is absolute motion, we find absolute motion inconceivable."

But why are we obliged to think of absolute motion, that is, of motion relative to no other existing thing whatever? It is plain we never think of it in the ordinary processes of human thought, but of relative motions alone. Our current ideas, judgments, and feelings about the motions of bodies are wholly unaffected by the question whether the whole solar system is in motion towards the constellation Hercules. And suppose this to be scientifically proved, still it is merely a new relation, added to those which were known before. If we affirm an absolute space, prior to and independent of all actual existence, then it would be natural to conceive also of an absolute motion. But if space be, as Mr Spencer argues, "the abstract of all the relations of coexistence among real things," and motion is change of place, that is, of position relative to other things, then an absolute motion, that is, relative to no existence whatever, may well be inconceivable, since it would contradict the definition. But surely no one, who holds the relativity of all knowledge for one main article of his creed, ought to say that the relativity of all motion is an absurdity and open contradiction.

So again of the change from Rest to Motion, or from Motion to Rest. It is surely a most extreme paradox to say that either of these is unthinkable. Mr Mill may well say, of such a style of philosophizing, that the number of its thinkable objects must be remarkably small. In point of fact, no conceptions are more usual, familiar, and unavoidable. If any finite change is inconceivable and impossible, because there is an infinite disproportion between any change at all and no change whatever, the whole of the later superstructure of Mr Spencer's philosophy comes to a sudden and calamitous end. Instead of the indestructibility of force and motion, the grand result will be the impossibility of any motion or change whatever, a universe frozen down into eternal rest and sameness, because no part of it can begin to move without being guilty of a logical inconceivability and contradiction.

But the remarks on Force are the strangest and the most paradoxical. "We are obliged to say that the antagonist molecular forces of attraction do not both vary inversely as the square of the distance, which is unthinkable; or else that matter does not possess that attribute of resistance, by which it is distinguished from empty space, which is absurd."

The latter half of the statement is plainly true. To deny that matter resists compression contradicts all experience, and puts an end to all physical science at a blow. But what shall we say of the other part? A wilder assertion, or more palpably and demonstrably untrue, was never made. It is assumed as self-evident that atoms can both attract and repel each other by the same law, that of the inverse square of the distance, and next, that it results from the laws of space, that they must both attract and repel by this law, and no other. Now what is here affirmed to be necessary and an *à priori* truth is strictly impossible. For either the attraction and repulsion are equal, or one is greater. If both equal at one distance, they are equal at all distances, and neutralize each other, and thus there is neither attraction nor repulsion. If one greater, say attraction, at one distance, it is greater at all distances, and there can be no real repulsion anywhere in the universe. And since on this hypothesis, that no other law of variation is possible from the nature of things, both attraction and repulsion cannot exist, the question must arise, Why should it be attraction rather than repulsion, or repulsion rather than attraction? and to this no answer can be given.

But besides this superlative error, the main assertion, that the law of gravitation is a mere corollary from the nature of space, is a monstrous inversion of the evident truth. The Principia are thus transformed, by a stroke of the pen, into one long process of laborious folly. They become an attempt to prove, by subtle geometrical arguments, and a comparison of their results with the facts of observation, what is self-evident from the laws of space. But physicists, we are told, are obliged to take this law, because the negation of it is inconceivable! Yet nearly one half of the First Book of the Principia is occupied with tracing out the curves answering to other possible laws of force, and laws of force needed for the description of various curves, in fixed and in moveable orbits. Similar theorems form a considerable part of almost every later work on theoretical dynamics. The law of gravitation may be viewed, either as primary and ultimate, or as an indirect result of some still unknown law of ethereal action, in connection with the geometrical laws of space. In the former view it must appear as one law chosen out of many alternatives, equally conceivable with itself, and referred directly, not to a fatal necessity, but the choice of the Supreme Intelligence. On the other view, the *à posteriori* element belongs to the other laws of the constitution of ether, and its action on matter, without which the geometrical laws of space would be utterly powerless to cause any local change or motion whatever.

The last subject, on which hopeless contradiction is affirmed to exist, is Consciousness and Personality. And the contradiction asserted is twofold. We cannot conceive in thought either the extent of consciousness, or its substance.

And first of its extent, we can neither conceive it, we are told, as infinite nor as finite. In plain words, we can neither think of ourselves as perishing at some future time, nor yet as living for ever. The annihilation of the soul and its immortality are both unthinkable. A startling doctrine! How is it attempted to be proved? I will quote the passage.

"It may be said, though we cannot know consciousness to be finite in duration, because neither of its limits can be actually revealed, yet we can very well conceive it to be so. No, not even this is true. We cannot conceive the terminations of that consciousness which alone we really know, our own, any more than we can *per*ceive its terminations. For the two acts are here one. In either case the terminations must not be presented in thought, but represented, and as in the act of occurring. Now to represent the termination of consciousness as occurring in ourselves, is to think of ourselves as contemplating the cessation of the last state of consciousness, and this implies a supposed continuance of consciousness after its last state, which is absurd. In the second place, if we study the phenomena as occurring in others, or in the abstract, we are equally foiled....A last state of consciousness, like any other, can exist only through a perception of its relations to previous states. But such a perception of its relations must constitute a state later than the last, which is a contradiction."

This is marvellous reasoning. Let us take a particular case, that we may see its exact nature more clearly. A commonplace unbeliever says, I have no faith in a life to come. I conceive that I shall live just to the end of the nineteenth century, and then my consciousness will cease. I shall die, and live no more. I shall have melted into the infinite azure of the past. The philosophy of the Unknowable replies—You deceive yourself. You can conceive nothing of the kind. To think of your consciousness as having ceased in the year 1901, is to think of yourself as still alive in that year, and thinking of your death, at the last hour of the previous December, as an accomplished event. Thus you suppose a continuance of living consciousness after your last moment of consciousness, which is absurd. But it is the reasoning which is absurd, and not the conception it pretends to prove impossible. To think *of* the twentieth century, and to be still alive and thinking *in* the twentieth century, are not the same, but two wholly different things.

The other argument is no less preposterous. We cannot conceive a last state of consciousness in another person, because this state can exist only by a perception of its relation to previous states, and this perception must constitute a later state than itself, which is a contradiction. But if there be any truth in such an argument, it must apply to every other state of consciousness. It will prove, not that there can be no ultimate state, but no state of consciousness whatever. For if the essence of any state of consciousness is the comparison of itself with former states, and this comparison constitutes a later state, then every state of consciousness must be later in time than itself, which is absurd. The absurdity, however, clearly belongs, not to the original conception, either of a last state of consciousness or any other state, but to the argument which is brought to prove them inconceivable.

The other argument refers to the conscious substance, the self which feels and thinks. Mr Spencer reasons as follows:

"Unavoidable as is this belief (in the reality of the individual mind), established though it is, not only by the assent of mankind at large, endorsed by divers philosophers, but by the suicide of the sceptical argument, it is yet a belief which reason, when pressed for a distinct answer, rejects. It may readily be shewn that a cognition of self, properly so called, is absolutely negatived by the laws of thought...The fundamental condition to all consciousness is the antithesis of subject and object. But now what is the corollary from this doctrine as bearing on the consciousness of self? The mental act in which self is known implies a perceiving subject and a perceived object. If then the object perceived is self, what is the subject that perceives? Or if it be the true self that thinks, what other self can it be that is thought of? Clearly a true cognition of self implies a state in which the knowing and the known are one, and this Mr Mansel rightly holds to be the annihilation of both. So that the personality of which each is conscious, is a thing which cannot truly be known at all. Knowledge of it is forbidden by the very nature of thought."

The sceptical argument seems here to possess remarkable vitality, after it has committed suicide, since it has power to discredit and disprove the most widely accepted of all truths, and to convict it of essential incongruity and self-contradiction.

"Self-knowledge is forbidden by the very nature of thought." This dictum stands in strange contrast to the opinion, so widely received among the old heathens of Greece and Rome, that the maxim, "Know thyself," came down from heaven, because it was the voice of higher than human wisdom. Which is of more value, the old lamp or the new? Let us examine the latter more closely.

Cognition of self implies a state in which the knowing and the known are one. That is certainly true. But one in what sense? A unity which excludes all duality or plurality? Clearly just the reverse. Our belief in a sentient, percipient mind involves the faith that the mind is one, while its sensations, impressions, perceptions, and beliefs are various and manifold. It is a unity, which does not exclude a plurality of thoughts and moments of existence, but implies and requires them. So our belief in the capacity of the mind to reflect on itself, to think on its own past or future states, is the belief in a unity which does not exclude but imply a certain duality. The statement, which Mr Spencer quotes with praise, and makes his fulcrum for uprooting all possibility of self-knowledge, is this:—If I think of myself, then I who think, and the self thought of, are both annihilated. Why, the exact reverse is self-evident, that if either the thinking self or the self thought of has no existence, the thought of self becomes non-existent and impossible. The existence of any sentient being implies unity in plurality, one and the same living thing, having still a multitude of successive feelings and sensations. The existence of any being capable of self-knowledge, or of the large class of thoughts which Locke calls ideas of reflection, equally implies unity in duality. The idea is no contradiction, though it may border on mystery. Even Mr Spencer will surely allow that we can form a conception of a circular line. But while such a circle is one figure, and always conceived as one, it is invariably conceived as having a convex and a concave side.

The doctrine of the whole chapter is summed up in these words:

"The explanation of that which is explicable does but bring out into greater clearness the inexplicableness of that which remains behind. The man of science sees himself in the midst of perpetual changes, of which he can discover neither the beginning nor the end...If he looks inward, both ends of the line of consciousness are beyond his grasp, nay, even beyond his power to think of as having existed, or as existing in time to come...Objective and subjective things he ascertains to be alike inscrutable in their substance and genesis. In all directions his investigations bring him face to face with an insoluble enigma, and he ever more clearly perceives it to be insoluble. He learns at once the greatness and the littleness of the human intellect; its power in dealing with all that comes within the range of experience, its impotence in dealing with all that transcends experience. He realizes with special vividness the incomprehensibleness of the simplest fact, considered in itself. He, more than any other, knows that in its ultimate essence nothing can be known."

The one element of truth in this passage has been long anticipated in those words of the Apostle—"If any man think that he knoweth anything," that is, fully and exhaustively, "he knoweth nothing yet as he ought to know." The sense of a deep, unfathomable mystery, accompanying and encompassing all human knowledge, has ever been most vividly felt, most fully and constantly recognized, by those whose wisdom, faith, and piety have been the most profound. But when this grand and simple truth of Scripture and universal reason is made the plea for a philosophy, which denies and discards every doctrine of religion, natural or revealed, and replaces all Theology by a theory of Physical Fatalism, the contradictions recoil, so as to crush and condemn that very system of thought they were intended to enthrone on the entire overthrow of religious faith. On the hypothesis now examined there can be no science, for God, Space, Time, Matter, Force, and Consciousness are all alike unthinkable and unknowable. There can be no man of science, for reason, when pressed to give answer, rejects the belief in the reality of any individual mind. If such a being could be conceived to exist, which has been denied, that he should see himself would be impossible. For he would thus be at once subject and object, and this "is rightly held to be the annihilation of both." He cannot see himself to be in the midst of changes, for a change from motion to rest, or from rest to motion, is inconceivable, and a change even in the rate or direction of motion must, for the very same reason, be impossible to be conceived. He may learn from such a theory the littleness of the human intellect, for what can be less than that which can know nothing, can even think of nothing, without falling into flat contradiction? But assuredly he can never learn its greatness. Far from teaching him the power of the human mind to deal with all that comes within the range of experience, the theory teaches him the exact reverse, that his power is impotence, and that from the merest atom of lifeless matter, up to the throne of the Almighty, human thought cannot make a single averment without proving itself blind and self-deceived. There is no room in the reasonings I have examined for any contrast between things that lie within the range of experience, and others that transcend it. Experience can be nothing more than a vain phantasmagoria of ever-fleeting shadows, if it be true of every object of human thought, of Matter, Mind, and the Great Author of the universe, that they are wholly unknowable and unknown. The whole course of false reasoning must be reversed, and the mystery which attends all partial knowledge must be carefully distinguished from unreality, self-contradiction, and falsehood, before there can be standing-ground, not only for Christian Theology, but for Physical Science itself. And the same course of thought, which breaks through these spiders' webs of false reasoning in physics, and shows that genuine sciences of space and time, of matter, motion and force, are all attainable, and have been attained, applies with equal emphasis to the highest subject of all, the knowledge of our Creator, whom to know in his works, as Newton has truly observed, should be one main purpose and aim of all genuine philosophy.

CHAPTER III.

ON THE RELATIVITY OF KNOWLEDGE.

THE Relativity of Knowledge is a doctrine which is very prominent in recent works of philosophy. Sir W. Hamilton, Dean Mansel, Mr Mill and Mr Spencer, with others almost of equal reputation, agree in affirming its truth and high importance. The last-named writer begins his exposition in these words:

"The conviction that human intelligence is incapable of absolute knowledge is one that has been slowly gaining ground, as civilization has advanced. Each new ontological theory, from time to time propounded, has been followed by a new criticism, leading to a new scepticism. All possible conceptions have been tried and found wanting; and so the entire field of speculation has been exhausted, without positive result, the only one arrived at being negative—that the Reality existing behind all appearances is and must ever be unknown. To this conclusion almost every thinker of note has subscribed. With the exception, says Sir W. Hamilton, of a few absolutist theorizers of Germany, this is the truth of all others most harmoniously re-echoed by every philosopher of every school. And among these he names Protagoras, Aristotle, Augustin, Boethius, Averroes, Albertus Magnus,

Gerson, Leo Hebraeus, Melanchthon, Scaliger, F. Piccolomini, Geordano Bruno, Campanella, Bacon, Spinoza, Newton and Kant."

Such a triple concurrence of the great names of past ages, the leading philosophers of the present day, and the whole course and drift of modern civilization, has a very imposing sound. It may well seem immodest to refuse or suspend our assent to a dogma, sustained by such a weight of concurrent authority. But here the principle itself comes to our aid, and restores to us some degree of liberty again. The *apparent* consent, Sir W. Hamilton and Mr Spencer assure us, is great and overwhelming. But what do we or can we know of the reality behind this appearance?

The doctrine is certainly left in a very paradoxical state by the recent discussions. Sir W. Hamilton, and his able disciple, Dean Mansel, stand foremost among its supposed advocates. Mr Spencer quotes many pages from both of them, as giving the clearest expression to the opinion he shares with them. But he then proceeds to reverse and disprove one main premise on which all their reasoning depends, that the notion of the Absolute, the Infinite, or the Unconditioned, is only negative. He maintains that it is positive, and even the most positive, persistent and real, of all ideas. Mr Mill, again, states that the doctrine is "true, fundamental, and of important consequences in philosophy." He makes it the common creed of Berkeley and the Sceptic Idealists, including Hume, of Kant and the Transcendentalists, of Hartley and the whole school of Mental Physiologists. But he remarks further that it may shade down "through a number of gradations, successively more thin and unsubstantial, till it fades into a truism leading to no consequences, and hardly worth enunciating in words." He proceeds to show, through several chapters, that Sir W. Hamilton, who affirms it strongly in one set of passages, contradicts it as strongly in another, and could only have held it in that sense, which reduces it most completely to a barren truism. And when Dean Mansel would disprove this charge by a different construction of the Hamiltonian Realism, Mr Mill and Prof. Fraser, by appeal to Sir W. Hamilton's own words, refute and disprove the attempted reconciliation.

Again, Mr Mill at first defines the doctrine, as held by Mr Spencer, to include two things, the certain existence of Things in themselves, and their absolute and eternal relegation to the region of the Unknowable. But he presently retracts his mistake (Exam. pp. 13, 181, 182); and while he thinks Mr Spencer's services to philosophy in the defence of the experience hypothesis are beyond all price, says that his real view recognizes as attainable "a prodigious amount of knowledge respecting the Unknowable." It must thus, in his view, be nearly as adverse as Sir W. Hamilton's Natural Realism to a genuine acceptance of the doctrine of Relativity in its own proper meaning.

But if the consent of its recent champions is phenomenal and illusive, so also is that of the earlier authorities. Mr Mill remarks as follows on the statement of Sir W. Hamilton, which Mr Spencer has taken for the basis of his own discussion :

"He supports his assertion by quotations from seventeen thinkers of eminence, beginning with Protagoras and Aristotle, and ending with Kant. Gladly, however, as I should learn that a philosophical truth, destructive of so great a mass of baseless and misleading speculations, had been universally recognized by philosophers of all past time, and that Ontology, instead of being, as I believed, the oldest form of philosophy, was a recent invention of Schelling and Hegel, I am obliged to confess that none of the passages, except the one from the elder Scaliger, and one from Newton, convey to my mind that the writers had ever come in sight of the great truth he supposes them to have intended to express. Almost all of them seem to me perfectly compatible with the rejection of it."

Once more Mr Spencer, after quoting with approval six pages from Sir W. Hamilton and Dean Mansel, in which they unfold this great doctrine, proceeds in the rest of the chapter to correct their supposed exaggeration, and expound in these words a wholly different view.

"The answer of pure Logic is held to be that, by the limits of our intelligence, we are rigorously confined within the relative, and that any thing transcending the relative can be thought of only as a pure negation, or as a non-existence. Unavoidable as this conclusion seems, it involves, I think, a grave error. If the premise be granted, the conclusion must doubtless be admitted; but the premise, in the form prescribed, is not strictly true. Though the arguments used by these writers to prove the Absolute unknowable have been approvingly quoted, and enforced by others equally thorough-going, there remains a qualification, which saves us from the scepticism otherwise necessitated. These propositions are imperfect statements of the truth, omitting or rather excluding an all-important fact. Besides that definite consciousness, of which logic formulates the laws, there is an indefinite consciousness, which cannot be formulated. Besides complete thoughts, there are thoughts which it is impossible to complete, and yet which are real, being normal affections of the intellect. Every one of the arguments, by which the relativity of our knowledge is demonstrated, postulates the positive existence of something beyond the relative. To say that we cannot know the Absolute is by implication to affirm that there is an Absolute. In the denial of our power to learn *what* it is, there lies hidden the assumption *that* it is; and this proves that the Absolute has been present to our minds, not as a nothing, but as a something. The Noumenon, everywhere named as the antithesis of the Phenomenon, is throughout necessarily thought of as an actuality. Our conception of the Relative itself disappears, if our conception of the Absolute is a pure negation."

Thus the doctrine of Relativity, that wide river, which confines us to a land of shadows, and shuts us out from any possible knowledge of "things in themselves," of Matter, Mind, God, in their true and real being, seems to part itself into four or five streams, and gives us some hope that we may be able to ford it, and pass over safely to the other side. Let us endeavour, with Mr Mill's aid, to discriminate these varieties.

The first and simplest is the doctrine that there is neither mind nor matter, Ego, or Non-ego, but sensations themselves alone, and is thus described:

"According to one of the forms, the sensations we are said in common parlance to receive from the objects are not only all that we can possibly know of them, but are all that we have any ground for believing to exist. What we term an object is but a complex conception, made up by the laws of association out of the ideas of various sensations, which we are accustomed to receive simultaneously. There is nothing real in the process but these sensations. They do not indeed succeed each other at random. They are held together by a law, that is, they occur in fixed groups, and in fixed order of succession: but we have no evidence of anything which, not being a sensation, is a substratum or hidden cause of sensations. That idea is a mental creation, to which we have no reason to think that there is any corresponding reality exterior to our minds. Such is the first and most extreme form of the doctrine. That is, not merely all we can possibly know of anything is the manner in which it affects the human faculties, but there is nothing else to be known; affections of human or other minds are all we can know to exist."

Next follows a second and more usual form of the doctrine, thus explained:

"The difference between the Ego and Non-ego is not one of language only, nor a formal distinction between two aspects of the same reality; but denotes two realities, each having a separate existence, and neither of them dependent on the other. They believe that there is a real universe of 'Things in themselves,' but as to what this Thing *is* in itself, we can only know what our senses tell us, and as they tell us only the impression it makes on *us*, we do not know what it is in itself at all. External things exist, and have an inmost nature, but their inmost nature is inaccessible to our faculties. If things have an inmost nature, apart not only from the impressions they produce, but from all they are fitted to produce, on any sentient being, this inmost nature is unknowable, inscrutable, and inconceivable, not to us only, but to every other creature. To say that even the Creator could know it, is to use language which to us has no meaning, because we have no faculties by which to apprehend that there is any such thing for him to know."

The position of Mr Mill himself seems to lie midway between these two forms of the doctrine. The first, as the simplest and purest, has his affections. With regard to matter, or the Non-ego, he adopts it fully. An orange, or a tree, or the city of Calcutta, is simply a group of "permanent possibilities of sensation." He accepts, like so many others, Berkeley's half of the phenomenal theory. And his love of consistency and logical completeness makes him look with a wistful longing on Hume's speculation, as its natural complement, that mind has no more existence than matter, and that the series of sensations, impressions, or feelings implies no person who feels, sees, or hears, and no thing tasted, felt, seen or heard. But Memory and Hope bar the path, and prevent him from acquiescing in this extreme conclusion. A remembrance, an expectation, is a present feeling, but involves a belief in more than its own present existence. This difficulty makes him, with great reluctance, leave one half of the theory ambiguous. That matter is only groups of sensations or possibilities of sensations is a settled conclusion of philosophy. That mind is only a series and sequence of feelings may be true also. The completeness of the doctrine of relativity seems to require it. The facts of hope and memory seem to exclude it. So that "the wisest thing we can do is to accept the inexplicable fact, without any theory of how it takes place (that is, whether or not there is a mind that hopes or remembers), and when we are obliged to speak of it in terms which assume a theory, to use them with a reservation as to their meaning." (Ex. pp. 247, 248.)

A fourth variety of the doctrine is that of Sir W. Hamilton. In many places he affirms it strongly, and without limitation. But he affirms as strongly elsewhere two doctrines, which constitute enormous exceptions to its completeness, and indeed spare so little as, in Mr Mill's view, to reduce it to a mere truism.

And first, in Physics he holds that secondary qualities are in the mind, but that the primary we directly cognize or discern in the objects themselves. This is his doctrine of Natural Realism, which he holds in contrast alike to Berkeley and the Sceptical Idealists, and to the large class of philosophers, whom he styles Cosmothetic Idealists, or Hypothetical Realists, and who believe in the existence of a material world, but only as an inference, and not a direct intuition.

The second main exception to the doctrine is in the higher field of religious thought. What has been excluded, as the very essence of the Philosophy of the Conditioned, under the name of knowledge, is brought back, at least in part, under the name of belief. "By a wonderful revelation," he says, "we are thus, in the very consciousness of our inability to conceive aught above the relative and the finite, inspired with a belief in a something unconditioned, beyond the sphere of all comprehensible reality." Thus the doctrine of relativity is doubly modified; with regard to Matter, by the doctrine of our direct cognition of its primary qualities, extension, resistance and form, as they are in matter itself; and in the sphere of Theology, by the assertion of an irrepressible belief, which assures us that something absolute and infinite exists, though our reason proves that the thought of it includes incredible contradictions.

The fifth and last form of the doctrine is Mr Spencer's own. It is introduced as an important correction of those statements of Sir W. Hamilton and Dean Mansel, which he has just before made the groundwork of his whole theory of the Unknowable. It is difficult, in these misty mountain-tops of thought, to be sure that we have rightly apprehended the exact meaning of an author, even after our best and most careful efforts. But the doctrine of Mr Spencer seems to me to include three successive courses of thought. In the first, he affirms the doctrine of relativity in its most extreme form, borrowing the arguments of Sir W. Hamilton and Dean Mansel, and enlarging them by an addition of his own. In the next, he reverses the argument, shows the falsity of its main premise, and makes the real existence of the Absolute the most valid and indestructible of all truths. In the third stage he retraces his steps once more, and attempts to remove the sense of illusion, which ordinarily follows the reading of metaphysics, and which he ascribes to the constant confusion of the popular and philosophical meaning of terms, by carrying out the idealist view one step further, and defining reality to be nothing more nor less than "persistence in consciousness." Thus he infers that the result must be the same to us, "whether what we perceive be the Unknowable itself, or an effect invariably wrought on us by the Unknowable." (F. P. pp. 160, 161.)

The first stage in this double alternation has been described already, and the chapter on relativity begins by repeating the arguments borrowed before from Sir W. Hamilton, with further additions. The sum of it may be given in two or three sentences. "The Absolute is a term expressing no object of thought, but only a denial of the relation by which thought is constituted.... To assume absolute existence as an object of thought is thus to suppose a relation existing when the related terms exist no longer.... This does not imply that the Absolute cannot exist, but it implies most certainly that we cannot conceive it as existing....The infinite, from a human point of view, is merely a name for the absence of those conditions under which thought is possible." (pp. 77—79.)

So far the argument is borrowed from Sir W. Hamilton and Dean Mansel, and rests on the purely negative nature of the idea of the Unconditioned, under what they hold to be two opposite modes or extremes, the Absolute and the Infinite. But Mr Spencer proceeds to add a third proof of the doctrine, from *likeness*, as well as relation and difference. He writes on it as follows:

"A cognition of the Real, as distinct from the Phenomenal, must conform to this law of cognition in general. The First Cause, the Infinite, the Absolute, to be known at all, must be classed. To be positively thought of, it must be thought of as such or such. Can it be like in kind to anything of which we have sensible experience? Obviously not. That which is uncaused cannot be assimilated to that which is caused, the two being antithetically opposed. The Infinite cannot be grouped along with something finite; since in being so grouped it must be regarded as non-infinite. It is impossible to put the Absolute in the same category with anything relative, so long as it is defined as that of which no necessary relation can be predicated. Is then the Actual, though unthinkable by classification with the Apparent, thinkable by classification with itself? This supposition is equally absurd with the other. It implies the plurality of the First Cause, the Infinite, and the Absolute, and this implication is contradictory....Thus from the very nature of thought, the relativity of our knowledge is inferable in three several ways. As we find by analyzing it, a thought involves relation, difference, likeness. Whatever does not present each of these does not admit of cognition. And hence we may say that the Unconditioned, as presenting none of them, is trebly unthinkable."

Here, on the face of the passage, the paradox and contradiction is striking and complete. We cannot think at all of Something, which we can express by three distinct names, the First Cause, the Infinite, and the Absolute, to say nothing of two others, the Real in contrast to the Phenomenal, the Actual in contrast to the Apparent; and can also, under each of these names, construct a triple argument, from relation, difference and likeness, to prove it unthinkable. And the paradox is even greater, when we pursue it into details. We cannot think of the First Cause as a cause, and so by likeness to other causes, nor by difference, that is, by contrast with a second cause or an effect. We cannot think of the First Cause, the Infinite, the Absolute, by plurality; when the very names are threefold, and plainly must denote either three Things, Persons, Beings, or three aspects and attributes of one and the same Thing, Person or Being. We cannot think of the Unconditioned by difference or contrast, when the name is senseless and mere gibberish, unless it denotes that which is in contrast to the Conditioned.

But the argument, in this first stage, while directly applied to Theology alone—since the First Cause, the Infinite, the Absolute, are abstract names for the Thrice Holy One, in whom Christians believe, and whom they worship and adore—really includes all Physics and Humanity in its comprehensive range. For the very same reasons of unknowableness have been shown to apply to Space, Time, Matter, Force, and the Conscious Mind. It is the Real in contrast to the Phenomenal, the Actual in contrast to the Apparent, which is pronounced trebly unthinkable and unknowable. The Noumenon or the Noumena, the Reality or the Realities, are all alike beyond the reach of human knowledge. We cannot even know whether they are many or one.

But now we come to a second stage, in which the pure and perfect doctrine of relativity is modified and almost reversed. A few sentences will be enough to show the importance of these fresh admissions.

"To say that we cannot know the Absolute is by implication to affirm that there *is* an Absolute. The Noumenon, everywhere named as the antithesis of the Phenomenon, is necessarily thought of as an actuality. It is impossible to conceive that our knowledge is a knowledge of Appearances only, without conceiving at the same time a Reality of which they are appearances, for appearance without reality is unthinkable. Truly to realize in thought any one of the propositions of which the argument consists, the Unconditioned must be represented as positive and not negative. An argument which assigns to a certain term a certain meaning, and ends in shewing that this term has no such meaning, is simply an elaborate suicide. Clearly, then, the demonstration that a definite consciousness of the Absolute is impossible to us unavoidably presupposes an indefinite consciousness of it."

"The error consists in assuming that consciousness contains nothing but limits and conditions, to the entire neglect of that which is limited and conditioned. It is forgotten that there is something which forms the raw material of definite thought, and remains after the definiteness which thinking gave to it has been destroyed.... We are conscious of the Relative as existence under conditions and limits. It is impossible that these can be thought of, apart from something to which they gave the form. The abstraction of these is by hypothesis the abstraction of them only. Consequently there must be a residuary consciousness of something which filled up their outline, and this indefinite something constitutes our consciousness of the Non-Relative or Absolute. Though we cannot give the consciousness any qualitative or quantitative expression, it is not less certain that it remains with us as a positive and indestructible element of thought.... If the Absolute is present in thought only as a mere negation, the relation between it and the Relative becomes unthinkable. And if this relation is unthinkable, there is the Relative itself unthinkable for want of its antithesis, whence results the disappearance of all thought whatever. The momentum of thought inevitably carries us beyond conditioned existence to unconditioned, and this ever persists in us as the body of a thought to which we can give no shape. Hence our firm belief in objective reality, a belief which metaphysical objections cannot for a moment shake....An ever-present sense of real existence is the very basis of our intelligence. And since the only possible measure of relative validity among our beliefs is the degree of their persistence in opposition to efforts made to change them, it follows that this which persists at all times, under all circumstances, and cannot cease till consciousness ceases, has the highest validity of any....It is impossible to get rid of the consciousness of an actuality lying beyond appearances, and from this impossibility results our indestructible belief in that actuality."

This doctrine, though offered as a correction of Sir W. Hamilton's theory, is really much the same as his "wonderful revelation, which inspires us with a belief in something Unconditioned, beyond the sphere of comprehensible reality." The phrase alone is different. With

Mr Spencer, all realities constitute the Unknowable, while the Knowable includes appearances alone. The contrast is twice expressed as that of the Real and the Phenomenal, the Actual and the Apparent. The irresistible belief, however, is the same. One writer styles it the belief in something beyond the sphere of comprehensible reality; the other, in a reality or realities, lying behind and beyond those appearances, which alone we can really know. In one, it is the fixed belief, by revelation, of something of which we cannot even think without contradiction. In the other, it is the indestructible belief in a reality we cannot formulate in thought, and which persists in our minds as the body of a thought without shape or form.

So far, the view is simply the acceptance of the second of the two forms of the doctrine of Relativity, as described by Mr Mill, and the rejection of the first. There are realities behind the appearances, or one grand reality, but of which nothing can be known. But in another place the return journey is carried further, so that Mr Mill accounts it an entire surrender of the main doctrine itself. The more or less coherent relations among any one's states of consciousness are said to be generated by experience of the more or less constant relations in something beyond our consciousness. This Mr Mill expounds as an affirmation "that for every proposition we can truly assert about the similitudes, successions, and coexistences of our states of consciousness, there is a corresponding similitude, succession, and coexistence, really obtaining among Noumena beyond our consciousness, and even that we can have experience of the same." And he expresses a natural surprise that so able an advocate and champion of the experience hypothesis should recognize "this prodigious amount of knowledge respecting the Unknowable." In fact, the doctrine of Relativity is thus reduced to a prosaic version of Plato's noble and vivid figure. We dwell in a cave, in a land of ever-moving shadows; but those shadows have fixed and definite relations, which we can detect by careful experience, to unseen realities on which they depend.

But when, in this second stage of the theory, we seem almost to have emerged from the cave, and obtained glimpses of the real and the actual, a third stage ensues, which throws us back into our dream-land once more. Reality receives a new definition, and means only "persistence in consciousness." A thought continues, or continually recurs, in spite of every effort to get rid of it, and this persistency of the thought is the only thing, the sole reality. The opening of ch. III. in the second Book, on the Knowable, is occupied with the exposition of this view. And the passage contains so frank an admission with regard to the usual effect, on plain minds, of most metaphysical theories, and so important a remark with regard to one main cause of confusion and error, that I shall venture to quote it at length. It will form a natural basis for all the further remarks which I have to make, in this and the next chapter, on this doctrine of Relativity. Mr Spencer writes as follows:

"That sceptical state of mind which the criticisms of philosophy usually produce, is, in great measure, caused by the misinterpretation of words. A sense of universal illusion ordinarily follows the reading of metaphysics, and is strong in proportion as the argument has appeared conclusive. This sense of universal illusion would probably never have arisen, had the terms used been always rightly construed. Unfortunately, these terms have by association acquired meanings quite different from those given them in philosophical discussions; and the ordinary meanings being unavoidably suggested, there results more or less of that dream-like idealism which is so incongruous with our instinctive convictions. The word *phenomenon*, and its equivalent word, *appearance*, are in great part to blame for this. In ordinary speech these are uniformly employed in reference to visual perceptions. Habit almost disables us from thinking of appearance except as something seen; and though *phenomena* has a more generalized meaning, we cannot rid it of associations with *appearance*, its verbal equivalent. When, therefore, Philosophy proves that our knowledge of the external world can be but phenomenal, when it concludes that the things of which we are conscious are appearances; it inevitably arouses in us the notion of an illusiveness, like that to which our visual perceptions are so liable in comparison with our tactual perceptions. Good pictures show us that the aspects of things may be very nearly simulated by colours on canvas. The looking-glass more distinctly proves how deceptive is sight, unverified by touch. And the frequent cases in which we misinterpret impressions made on our eyes, and think we see something we do not see, further shake our faith in vision. So that the implication of uncertainty has infected the very word, *appearance*. Hence Philosophy, giving it an extended meaning, leads us to think of all our senses deceiving us in the same way that the eyes do, and so makes us feel ourselves floating in a world of phantasms. Did we in the place of them use the term, *effect*, which is equally applicable to all impressions made on consciousness through any of the senses, and carries with it in thought the necessary correlative *cause*, with which it is equally real, we should be in little danger of falling into the insanities of idealism."

"Such danger as might still remain would disappear on making a further verbal correction.... We increase the seeming unreality of that phenomenal existence which alone we can know, by contrasting it with a noumenal existence which we imagine would, if we could know it, be more real to us. But we delude ourselves with a verbal fiction. What is the meaning of the word *real*? This question underlies every metaphysical inquiry, and the neglect of it is the remaining cause of the chronic antagonisms of metaphysicians. In the interpretation of it, the discussions of philosophy retain one element of the vulgar conception of things, while they reject all its other elements, and create confusion by the inconsistency.

"The peasant, on contemplating an object, does not regard that which he contemplates as something in himself, but believes the thing of which he is conscious to be the external object, imagines that his consciousness extends to the very place where the object lies. To him the appearance and the reality are one and the same thing. The metaphysician, however, is convinced that the consciousness cannot embrace the reality, but only the appearance of it; and so he transfers the appearance into consciousness, and leaves the reality outside. This reality, left outside of consciousness, he continues to think of much in the same way as the ignorant man thinks of the appearance. Though the reality is asserted to be out of consciousness, yet the *realness* ascribed to it is constantly spoken of as though it were a knowledge possessed apart from consciousness. It seems to be forgotten that the conception of reality can be nothing more than some mode of consciousness; and the question to be considered is, What is the relation between this mode and others?

"By reality we mean persistence in consciousness; a

persistence that is either unconditional, as our consciousness of space, or conditional, as our consciousness of a body while grasping it. The real, as we conceive it, is distinguished solely by the test of persistence; for by this test we separate it from what we call the unreal. Between a person standing before us, and the idea of such a person, we discriminate by our ability to expel the idea from consciousness, and our inability, while looking at him, to expel the person from consciousness....How truly persistence is what we mean by reality is shewn in the fact that when, after criticism has proved that the real as we are conscious of it is not the objectively real, the indefinite notion we form of the objectively real is of something which persists absolutely, under all changes of mode, form, or appearance. The fact that we cannot form even an indefinite notion of the absolutely real, except as the absolutely persistent, clearly implies that persistence is our ultimate test of the real as present to consciousness."

"Reality, then, as we think it, being nothing more than persistence in consciousness, the result must be the same to us whether that which we perceive be the Unknowable itself, or an effect invariably wrought in us by the Unknowable. If, under constant conditions furnished by our constitution, some Power of which the nature is beyond conception always produces some mode of consciousness; if this mode of consciousness is as persistent as would be this Power, were it in consciousness, the reality will be to consciousness as complete in one case as the other. Were unconditioned Being itself present in thought, it could but be persistent, and if there is present Being, conditioned by the form of thought, but no less persistent, it must be to us no less real....Though reality under the forms of our consciousness be but a conditioned effect of the absolute reality, yet this conditioned effect standing in indissoluble relation with its unconditioned cause, and being equally persistent so long as the conditions persist, is to the consciousness supplying those conditions equally real. The persistent impressions, being persistent results of a persistent cause, are for practical purposes the same as the cause itself, and may habitually be dealt with as its equivalents....We deal with these relative realities as though they were absolutes, instead of effects of the absolute. And we may legitimately so do, as long as the conclusions to which they help us are understood as relative realities and not as absolute ones."

"That the relative reality answers to some absolute reality it is needful only for form's sake to assert. What has been said respecting the Unknown Cause, which produces in us the effects called Matter, Space and Time, will apply, on simply changing the terms, to Motion.... Force, as we know it, can be regarded only as a certain conditioned effect of the Unconditioned Cause, as the relative reality, indicating to us an Absolute Reality by which it is immediately produced. And here we see how inevitable is that transfigured realism to which sceptical criticism finally brings us round....Noumenon and phenomenon are presented as in their primordial relation, two sides of the same change, of which we are obliged to regard the last as no less real than the first."

The first paragraph of the above extract suggests a grave inquiry, which I must reserve for later consideration. If the criticisms of Philosophy usually produce a sceptical state of mind, and the current metaphysical theories issue commonly in "a sense of universal illusion," this is surely a strong presumption that Philosophy and Metaphysics have wandered into some deceptive by-paths, and forsaken the highway of Nature and Truth. The evil must probably be deeper and extend more widely than the ambiguous use of the words, *phenomenon* and *appearance*. The proposed remedy, however, seems to be only an aggravation of the disease. The sense of universal illusion is to be dispelled by giving a new sense to the word *reality*. We are to understand that its true meaning is only "persistence in consciousness." A thing or person is to be nothing more than our continued thinking, with no object of our thought, but the act or habit of thinking alone. The conception of reality can be nothing more than some mode of consciousness. Doubtless our thinking of a thing as real must be a mode of thought. But is the thing thought of a mode of thought? Is it not plain that to define reality in this way is exactly to reverse its true meaning? So that this ripest effort to remedy the illusiveness of metaphysics consists in affirming that things are nothing else than continued thinking, with no object of thought, and that Reality means only persistent non-reality.

Let us now compare together these three primary courses of reasoning, which constitute the foundation on which the new philosophy, the temple of Physical Fatalism, is reared. First, the ideas of the First Cause, the Absolute, the Infinite, and also of Space, Time, Matter, Motion, Force, Conscious Mind, are all contradictory and unthinkable. Atheism, Pantheism, and Theism are unthinkable. Space, finite or infinite, objective or subjective, is unthinkable. Matter, whether as a plenum, or discrete, finite atoms, or centres of force, is unthinkable. Motion cannot be thought of, either as absolute, or only relative. Force cannot be thought of, either as acting or not acting at a distance, as the same with matter, or distinct from it. Conscious mind cannot be conceived either as finite or infinite in duration, and knowledge of it is forbidden by the very nature of thought. Physics, Psychology and Theology are thus alike impossible sciences. We may well accept Mr Mill's comment, when replying to Sir W. Hamilton and Dean Mansel on their doctrine of the Infinite:—"We are unable, then, without contradiction, to assert anything not only of God, but of Time and of Space. Considered as a *reductio ad absurdum*, this is sufficient."

But now follows the first correction of the theory. We have an irresistible, indestructible belief in the Absolute, the Real, though all knowledge *what* it is, is forbidden by the very nature of thought. The Relative implies the Absolute, the Apparent implies the Real. "The momentum of thought carries us inevitably beyond conditioned existence to unconditioned; and this ever persists in us as the body of a thought, to which we can give no shape." "The conditioned effect stands in indissoluble relation to its unconditioned cause."

Here, then, we have, on one side, appearances which make up the Knowable, and realities which make up the Unknowable. We know that they are, by an irresistible belief, which no metaphysical criticisms can shake for a moment; but what they are remains, and must remain for ever, wholly unknown. Yet these unknown realities stand in indissoluble connection with the known appearances, and distinct sets of appearances represent and correspond with their own realities. I grasp a ball or an orange, I gaze upon a tree. The sensations of touch, of sight, are all that I can know. But I believe irresistibly that there is something I touch, something I look upon, and that it is I, who look, who touch, and reflect on my own past sensations. I can know nothing of these realities. But I may know that the orange phenomena stand in indissoluble connection with the real orange, the tree phenomena in like connection with the real tree, the mental phenomena with the real mind or Ego, and that the realities thus correspond with the sets of experiences or appearances, are represented by them, and for all practical purposes are equivalent to them. We may thus obtain, to borrow Mr Mill's remark, "a prodigious amount of knowledge respecting the Unknowable." The very same substitution, which in Theology has been rejected with scorn and indignation, as a new instance of the impiety of the pious, the attempt, when all speculative truth has been pronounced impossible, to supply the void with regulative truths, is at once accepted for the groundwork of a vast and imposing structure of Physical Science. In strict logic we can know nothing whatever of things in space and time, of matter, motion, force or mind. But we can know an immense deal about appearances and sensations, and may safely believe and act on the faith that the unknown realities exactly correspond in some way or other with these known appearances, though still we are bound to remember that all we really know belongs not to the realities but to the shadows alone.

But we have not yet reached the extreme of inconsistency and self-contradiction. The first stage of thought is that sensations, acts, or states of consciousness alone are knowable, and that things or persons, or the great First Cause, the Power which is behind the phenomena of the universe, are all unthinkable, and unknowable. The second stage is that the appearances suggest realities, and cannot be thought of without them; that belief in the realities is irresistible, though we cannot formulate this belief into clear knowledge, and that we still may know of a correspondence between the representing appearances and the realities represented by them. But in the third stage the scene shifts once more. We have no longer a world of unknown realities on one side and known appearances on the other, but two sets or kinds of realities, the relative and the absolute. "The relative reality answers to the absolute reality, and indicates an absolute reality, by which it is immediately produced." And reality itself has changed its meaning. It is no longer the fact of things or persons without us and distinct from us, and of our own existence, as distinct from a series of sensations, but persistency in consciousness. I have a mingled sensation, for instance, of round shape and yellow colour. If this sensation persists and never intermits, it is a real orange, and the only real orange. If it only fluctuates and comes and goes uncertainly, it is an unreal orange. Persistent vivid manifestations of something unknowable, which we usually style sensations, are real matter. Persistent faint manifestations of the unknowable something, which answer to what are usually called reflections, are the real Ego or mind. The irresistible belief in the objective reality of matter in the world of space around us, and of conscious mind, our own, and that of other men, after being affirmed in the strongest terms, is contradicted and set aside once more. There are no real objects of thought without us, for this would be to admit something outside of consciousness. We, who think and feel, have no real existence. But some thoughts or sets of thoughts, some sensations or sets of sensations, last longer, or recur more obstinately than the rest; and these, if they are vivid, are the only real matter, and if they are faint and shadowy, are the only real mind. And both alike, whether vivid and clear, or faint and shadowy, agree in this one character, that they are manifestations of the Unknowable. The only realities are transient, momentary glimpses, often repeated, of something or other which can never be seen, manifestations of one or many Unknowables, of which we cannot tell whether it be one or many, and which it is impossible, by the laws of thought, that it can be manifested or known.

Such is the doctrine of the Relativity of Knowledge, as taught by Mr Spencer in his 'First Principles of Philosophy,' and in the shape it finally assumes. It illustrates the main idea of the Fatalistic Theory of the Universe, which is presently built upon it, and which looks upon the whole course of physical change as one immense series of oscillations of alternate evolution and dissolution. It begins with an intellectual chaos, in which every possible subject of thought is pronounced unthinkable, and consigned to mist and eternal darkness. It oscillates into some dawn of light and science, when it recognizes our irresistible belief in realities, and that appearances so truly represent these realities, as to lead to a large practical knowledge of all that theory has styled unknowable. And then it oscillates once more into the hypothesis, which denies all reality but a mere counterfeit, and turns the whole world of science into a series of roads that start from nothing and lead to nothing, thoughts more or less continuous, which no real person thinks, and in which no real thing or person is thought of. In this form I shall venture to apply to it Mr Mill's own stricture on Sir W. Hamilton's "Law of the Conditioned," a still more famous variety of the same general view. It "must be placed in that numerous class of metaphysical doctrines, which have a magnificent sound, but are empty of the smallest substance."

CHAPTER IV.

THE RELATIVITY OF KNOWLEDGE ACCORDING TO SIR W. HAMILTON AND MR MILL.

THE Relativity of Knowledge, I have shewn in the last chapter, has a different meaning in each of the three philosophers, whose seeming verbal agreement, and high reputation, are at first sight a strong presumption in favour of its truth. Sir W. Hamilton, Mr Spencer, Mr Mill, all directly affirm it in the most emphatic terms. The last of them pronounces it to be true, fundamental, and full of important consequences. Sir W. Hamilton, whose immense reading in metaphysics is well known, claims for it the general assent of all great thinkers of former days, and speaks of a few German theorizers, such as Schelling and Hegel, as the only exceptions. Mr Spencer adopts from him both the historical statement, and the exposition of the doctrine, and places it at the foundation of his own laboriously constructed scheme of philosophy. But when we look more closely, we find that Mr Mill reduces Sir W. Hamilton's seventeen authorities to two only, though he would gladly have learned that there had been so wide an acceptance of what he calls "a philosophical truth destructive of a great mass of misleading speculation;" and of these two, Newton, the more weighty, may be shewn, by the context of the quotation, to be as far as the others from affirming the real doctrine in debate. Also Mr Mill proceeds to prove, through six or seven chapters, that the Natural Realism of Sir W. Hamilton, and his doctrine of Belief, as distinct from knowledge, amount to a practical surrender and reversal of the doctrine of Relativity, which in words he so strongly affirms. In like manner he corrects his first impressions of Mr Spencer's view, and concludes that it is really no less inconsistent than the Scotch philosopher's with the genuine doctrine, since it recognizes as attainable "a prodigious amount of knowledge of the Unknowable." The weight of concurrent authority being thus proved a mere shadow, the way is open for a direct inquiry into the truth of the doctrine, as held by Mr Mill himself, in what he conceives to be its proper and only consistent form.

The Phenomenalism, derived from Berkeley, Hume, and Kant, which has prevailed so widely in nearly all later metaphysics, and which Mr Mill undertakes to expound, is open, at the outset, to two weighty grounds of suspicion. The first is admitted by Mr Spencer, in the statement before quoted, that metaphysics of this type "usually produce a sceptical state of mind" and are ordinarily followed by "a sense of universal illusion." No such result follows the genuine discoveries of science in other fields of thought. No sense of illusion haunts the simplest reader, when introduced to the writings of Bacon or Newton, of Herschel, Cuvier, Davy or Faraday, of practical moralists or of Christian divines. He ascribes the feeling to an ambiguous use of words, especially phenomenon, appearance, and reality. But such an effect must surely have a wider and deeper cause than the faulty use of two or three words, while the proposed remedy aggravates the disease. It is to define reality as "persistence in consciousness," a definition truly unthinkable, and never known or heard of till the 'First Principles' appeared.

The second ground of suspicion is that the doctrine runs counter to the laws and habits of human speech in every known dialect. It might thus seem almost disproved by the history of language alone. It teaches that feelings, sensations, impressions or ideas are the only proper objects of knowledge, and that substances, things, or persons, are either mere mental illusions, or wholly unknowable and unknown. The Realism which holds that we perceive the primary qualities of matter in the things themselves, and the hypothesis of an unformed half knowledge of the absolute realities, corresponding with our experiences of the relatives or appearances, are rejected by Mr Mill, as inconsistent with a genuine recognition of Relativity. But he complains, in his Logic, of the perverseness with which language takes every word denoting real existence, such as substance, thing, being, essence, entity, and applies them to what the doctrine pronounces unknowable, to material objects, minds, and persons, instead of sensations or states of consciousness, which it accounts the only knowable things. "Our sensations," he says, "seldom receive separate names." We have a name for the object, and for the quality, but for the sensation itself "language has provided us with no single-worded or immediate designation." Thus, if the doctrine be true, a strange perverseness has marked the growth, not only of the English tongue, but of every language known to history. Names expressive of real existence have been applied exclusively to things either non-existent, or wholly unknowable, while the only things that can be really known have either had to borrow titles from something else, or been left without a name.

The misuse of words, I agree with Mr Spencer, has much to do with that sense of illusion, which commonly attends the reading of many metaphysical works. But I believe that the cause lies deeper than the ambiguity in the terms, phenomenon, appearance, and reality, and depends on a fundamental falsehood. Phenomenalism, in my opinion, is a faulty analysis of human thought and experience, followed by a synthesis no less faulty. Philosophers or metaphysicians, in the name of science, try to unlearn and unteach truths, which they, as well as the peasant, have been learning from the very first moment of conscious life, and after all their efforts can never really unlearn. As a speculation, the non-reality of matter, or even of mind, may be patched upon the garment of their habitual course of thought. But it cannot and will not cohere. As soon as the violent effort ceases, truth resumes its settled power, and the illusion melts away of its own accord. The philosopher, like the peasant, is compelled to accept the lesson, taught by his own experience, and the experience of all the rest of mankind. He knows and feels, whatever speculators may say, that he is a real person, living, acting, thinking, moving, in the midst of a real world of outward things.

Hume, one of the ablest sceptics, to whom we owe the proposed extension of Berkeley's reasoning on the non-existence of matter to mind also, has frankly acknowledged this result of the theory. "These principles," he says, "may flourish and triumph in the schools, where it is difficult, if not impossible, to refute them. But as soon as they leave the shade, and by the presence of the real objects are put in opposition to the more powerful principles of our nature, they vanish like smoke, and leave the most determined sceptic in the same condition as other mortals."

The doctrine I have to examine has been variously stated as follows. First, by Sir W. Hamilton: "Of things absolutely or in themselves, be they internal or external, we know nothing, or know them only as incognizable. We become aware of their incomprehensible existence, only as this is incidentally and accidentally revealed to us through certain qualities related to our faculties, and which qualities we cannot think as unconditioned and irrelative. All that we know is phenomenal, phenomenal of the unknown." By Mr Mill in his Logic: "It may safely be laid down as a truth, obvious in itself, and admitted by all whom it is at present necessary to take into consideration, that of the outward world we know and can know absolutely nothing, except the sensations we experience from it." Sir W. Hamilton once more, Lect. I. 137: "This something, absolutely and in itself, apart from its phenomena, is to us zero. It is only in its qualities, in its effects, in its relative or phenomenal existence, that it is cognizable or conceivable....Matter or material substance, as contradistinguished from these qualities, is the name of something unknown and inconceivable. The same is true with regard to the term, mind. Our whole knowledge of mind or matter is thus only relative. Of existence, absolutely and in itself we know nothing; and we may say of man what Virgil said of Æneas, contemplating in the prophetic sculpture of his shield the future glories of Rome—
Rerumque ignarus, imagine gaudet."

Here it is plainly taught that all our knowledge is imaginary and non-real, of mere shadows and images, and not of things themselves. The sense of illusion it awakens, then, in common minds, is no misconception arising from an ambiguous use of the words *phenomenon* and *appearance*. On the contrary, it springs out of the very definition of the doctrine of Relativity, which Sir W. Hamilton has laid down, and which Mr Mill approves. The philosophers of this school do "make us feel ourselves floating in a world of phantasms." Man, it says, is ignorant of realities, and rejoices in shadowy images alone.

To see the nature of the doctrine more clearly, I shall quote a passage from Sir W. Hamilton, which Mr Mill highly approves, as shewing that he had a greater capacity for the subject than many metaphysicians of high reputation, and particularly than his predecessors, Reid and Stewart. I believe, on the contrary, that Sir W. Hamilton, Mr Spencer, Mr Mill, and all the phenomenalists, have here at the outset gone wrong together. The two first mitigate the original fault by admissions opposed to their common premise, but agreeing with truth and common sense; while Mr Mill, more logically consistent, is thereby led, from his false premise, still deeper into error. Sir W. Hamilton writes as follows:—

"A fact of consciousness is that whose existence is guaranteed by an original and necessary belief. But there is an important distinction to be made, which has been overlooked by all philosophers, and led some of the most distinguished into no inconsiderable errors.

"The facts of consciousness are to be considered in two points of view, either as evidencing their own phenomenal existence, or the objective existence of something else beyond them. A belief in the former is not identical with a belief in the latter. The one cannot, the other may possibly be refuted. In the case of a common witness, we cannot doubt the fact of his testimony, but we can always doubt the truth of that which his testimony avers. So it is with consciousness....In the act of External Perception, consciousness gives us a conjunct fact, the existence of Me or Self as perceiving, and the existence of something different from Me or Self, as perceived. Now the reality of this, as a subjective datum, as an ideal phenomenon, it is impossible to doubt without doubting the existence of consciousness; for consciousness is itself this fact, and to doubt the existence of consciousness is absolutely impossible. For as such a doubt could not exist, except in and through consciousness, it would consequently annihilate itself. We should doubt that we doubted. As contained in an act of consciousness, the contrast of mind knowing and matter known cannot be denied."

"But the whole phenomenon in consciousness may be admitted, and its inference disputed. Consciousness, it may be said, is only a phenomenon. The contrast between the subject and the object may be only apparent, not real. The object given as an external reality may be only a mental representation, which the mind is, by an unknown law, determined unconsciously to produce, and to mistake for something different from itself. All this may be said and believed without self-contradiction; nay, all this has, by the immense majority of modern philosophers, been said and believed."

"The assertion that the present existence of the phenomena of consciousness, and the reality of that to which these phenomena bear witness, rest on a foundation equally solid, is wholly untenable. The second fact, testified to, may be worthy of all credit, as I agree with Mr Stewart that it is; but still it does not rest on a foundation equally solid as the fact of the testimony itself. Mr Stewart confesses that of the first no doubt had ever been suggested by the boldest sceptic; and the latter, so far as it assures us of our having an immediate knowledge of the external world, has been doubted, nay even denied, not merely by sceptics, but by modern philosophers almost to a man."

"We are immediately conscious in perception of an ego and a non-ego known together, and in contrast to each other. This is the fact of the Duality of Consciousness. It is clear and manifest. When I concentrate my attention in the simplest act of perception, I return with the most irresistible conviction of two facts, or two branches of the same fact, that I am, and that something different from me exists. In this act I am conscious of myself as the perceiving subject, and of an external reality as the object perceived, and I am conscious of both existences in the same individual moment of intuition. The knowledge of the subject does not precede or follow the knowledge of the object; neither determines nor is determined by the other. Such is the fact of perception revealed in consciousness, and as it determines mankind in general to their almost equal assurance of the existence of an external world, as of the existence of our own minds. We may lay it down, as an undisputed truth, that consciousness gives, as an ultimate fact, a knowledge of the ego in contrast to the non-ego, and a knowledge of the non-ego in contrast to the ego. Again, consciousness not only gives us a duality, but gives its elements in equal counterpoise and independence. The ego and non-ego, mind and matter, are not only given together, but in absolute co-equality. The one does not precede, the other does not follow. Each is equally dependent, equally independent. Such is the fact as given in and by consciousness....Philosophers, however, have not been content to accept the fact in its integrity, but to accept it only under such modifications as it suited their systems to devise. In truth there are just as many different systems originating in this fact, as it admits of various possible modifications. There is first the grand division of philosophers into those who do and do not accept the fact in its integrity. Of modern philosophers, almost all are comprehended under the latter category; while of the former, if we do not remount to the schoolmen and the ancients, I am only aware of a single philosopher, before Reid, who did not reject, in part at least, the fact as consciousness affords it. As it is expedient to possess a precise name for a precise distinction, I would denominate those who implicitly acquiesce in the primitive duality, as given in consciousness, Natural Realists or Natural Dualists."

The doctrine of Sir W. Hamilton may thus be summed up in the following propositions. Consciousness is a general term for every mental act or impression, so that all mental philosophy is made up of facts or states of consciousness. Every act of sensation or perception is essentially triune. There is the phenomenal fact of consciousness, which cannot be disputed or denied. There is the double affirmation of a subject or ego, and of an object or non-ego, which cannot be denied as a fact; but the two things affirmed may be, and often are, denied. The phenomenal act of perception is compared to a witness. It testifies to itself, and this first testimony is indisputable. Doubt here annihilates itself. It testifies to a real subject and a real object, and both at the same moment. And here its veracity may be denied, though it ought not to be. The certainty of the self-testimony is perfect and cannot be exceeded. The certainty of the double testimony to a subject and an object is not so great, since the denial is not a self-contradiction. But still it is a great error. It involves the consequence that our faculties have been given only to deceive us, and thus implies a denial of the Divine veracity.

Mr Mill adopts the same definition of consciousness, as a generic name for mental acts, impressions, or modifications of every kind. He accepts the contrast drawn between the phenomenal fact of consciousness, as self-testified and undeniable, and its testimony to a subject and an object, and counts it a proof of Sir W. Hamilton's special acuteness and metaphysical sagacity. He admits also the present testimony of consciousness to a subject and an outward object. But he thinks that an appeal lies from present to primitive consciousness; that the testimony is not immediate and intuitive, but acquired through repeated experiences, and certain laws of association; that, so far as it refers to material substance, it is illusive, a superfluous mental fiction, and may be replaced by a doctrine, which assumes no useless hypothesis, that of Permanent Possibilities of Sensation. His theory, to be complete, should adopt the view of Hume as well as that of Berkeley, and replace minds or persons by Permanent Capabilities or Passive Possibilities of Sensation, and Active Possibilities of Volition. But the facts of memory and hope are, he owns, inexplicable on this view, which he therefore cannot fully adopt, and yet will not wholly abandon. He submits, then, to use all the terms which imply the existence of mind as the unit to which all passing states of consciousness belong, but "with a reservation as to their meaning;" so that, after all, the existence of mind, like that of matter, may perhaps be an illusive fiction and nothing more.

Now I believe that the truth will be found in reversing nearly every one of these statements. First, the definition of consciousness, which both writers hold in common, is misleading and untrue. Next, the contrast which Sir W. Hamilton draws between the fact of the testimony of consciousness, which cannot be doubted, and its veracity, which may be denied—a contrast which, he says, has been overlooked by all previous philosophers, but which Mr Mill affirms to be unquestionably just, and a proof of his rival's great capacity for metaphysical research—is, I believe, not only untrue, but an illusive self-contradiction. Where they diverge from each other, I believe that each holds a partial truth, but mingles it with a distinct variety of error. Sir W. Hamilton is right in holding that there now accompanies every perceptive act a strong conviction of the reality of an outward thing perceived, and of our own mind which perceives it. But though the connection is so close that it may be called intuitive, as indeed its type is the impression made on us in beholding a landscape, I think that it is neither immediate nor simultaneous; that the impression of an object is direct, and precedes the reflex conviction of our own existence; and that each impression or conviction, as it now arises, is not simple and undivided, but a compound result of many past experiences, by which the conception of the object on one side, and the subject mind on the other, has been successively enriched and also defined. Mr Mill, I believe, is right when he affirms, in opposition to Sir W. Hamilton, that our present notions both of subject and object, are not simple and primitive data, given in one perceptive act, but are a composite result from many such acts, and what he styles "laws of inseparable association." But he is wrong in holding that the notion of material substance is a mere illusion, and not the proper and necessary result of the elementary act of perception, when many such have occurred, and their results have been combined. He is wholly wrong in supposing that we can have a firmer and clearer knowledge of sensations than of objects perceived; or that a sensation can be conceived to exist, without any perceiving mind; or that they can, as sensations, be grouped by laws, when things perceived and minds perceiving are both denied; or that there can be possibilities when all realities are denied; or that a permanent possibility of sensation can be more than an empty phrase without meaning, in the absence of any being that feels and perceives, and any being or existence that is a cause of sensation. It is like the shadow of a suspension bridge, floating in the air, with no support at either end, a line of possible transition from nothing to nothing.

I. And first, what is the proper meaning of consciousness? A kind of dimness and ambiguity in the use of this frequent word seems to me the first beginning of the whole amount of error in each of the two rival philosophies.

Consciousness, according to Reid, is an operation of the understanding, distinct from memory, the objects of which are "our pains, hopes, fears, desires, doubts, thoughts of every kind, all the passions, actions, and operations of our own minds, while they are present." Dr Brown, in his eleventh Lecture, rejects this definition. It erroneously doubles, he thinks, every mental act or impression. Consciousness, in his view, is only a generic name for the feelings or mental acts or impressions themselves. I see a tree, and I am conscious that I see a tree, are the same thought differently expressed. But he admits that the word may also be used to express a brief and rapid retrospect of feelings, perceptions, and emotions, so recent that the interval is overlooked and forgotten. Only these ought, he thinks, in strictness to be classed with other acts of memory.

Dr Brown's correction of Reid is accepted as true both by Mr Mill and Sir W. Hamilton. Consciousness they both treat as a generic term for every mental modification whatever. The history of a mind is thus a series of "states of consciousness." But here they diverge. Sir W. Hamilton affirms strongly that an objective reality is directly given or affirmed in each act of perception. Thus consciousness, with him, includes an outward object, as well as the mental modification in the act of perceiving; and the strange phrases are introduced, that we are conscious of the sun and the moon when we look upon them, and conscious of an apple, orange, rose, lump of iron, or a piece of gold or silver, when we see it, taste it, smell it, or touch it, and hold it in our hand. On the other hand, Mr Mill restricts the word to its popular use, when it refers to the mind only, or a modification of thought or feeling. But then, as the natural result of making it the genus of all mental acts or emotions, the direct perception of any thing without the mind is precluded and made impossible. The mind becomes its own prison. All its acts are acts of consciousness, and therefore knowledge of any thing beyond its own states becomes impossible.

But here language, with which metaphysical writers are too apt to play strange tricks, and overlook its plainest lessons, comes to our aid, and detects, I think, the root of the error. *Scire* and *conscire* are not the same. The first is simple, the second is compound. The first refers to direct and simple knowledge of any object. The second refers to recognition of the mind or inner self, along with some other object of the mind's knowledge.

It is an error to think, with Reid, that every mental act or impression is double, that I see a tree, or book, or house, and think, in the same indivisible point of time, that I am seeing and looking upon it. It is an error in Brown, who rightly rejects this duality, to assume that consciousness can be a fit and proper name for all mental acts and impressions whatever. It does not apply, except by an abuse of language, to direct sensations or perceptions. It applies only to the second main class of mental modifications, or what Locke styles ideas of reflection, when the mind, along with some other object of thought, is thinking of itself also. The alternative, which Dr Brown states clearly, but then passes by, gives the only proper use of the term. "When the retrospect is of very recent feelings,...the short interval is forgotten, and we think that the primary feeling, and our consideration of the feeling, are simultaneous....When consciousness is any thing more than the sensation, thought, or emotion, of which we are said to be conscious, it is a brief and rapid retrospect. Its object is not a present feeling, but a past feeling, as truly as when we look back, not on the moment immediately preceding, but on some distant event or emotion of boyhood."

One correction or extension only needs to be added to this remark. In the case of a momentary sensation or perception, consciousness is not a simultaneous, but an immediately sequent act of reflection. But we commonly mean by a sensation or perception, not a momentary act, but a state continued through many moments, and thus including a hundred or a thousand successive acts or impressions. It is plainly possible for reflection to alternate with observation; so that each of fifty moments of sensation may be followed by one of conscious reflection, and yet the whole may form one mingled state of sensation and consciousness, in which the two elements seem joined inseparably, like a piece of a hundred lines in rhyming couplets, where the odd and even lines, those which introduce a new rhyme, and those which repeat it, blend into one series, and are practically inseparable.

This faulty use of consciousness, extending it to all mental acts or states, is counteracted, in Sir W. Hamilton, by his doctrine of Natural Realism, or the direct apprehension by the mind of outward objects. Its chief result is verbal, in a barbarous phraseology, that we are conscious of an apple, an orange, a house or a tree, a friend or an enemy, instead of our own feelings, sensations or emotions concerning them. In Mr Mill the error goes much further, and shuts up the mind within its own states and feelings, as a prison-house, beyond which it can see and know nothing. As Mr Spencer has well observed, in his article on the Test of Truth: "If we decline to acknowledge anything beyond consciousness till it is proved, we may go on reasoning for ever, without getting any further, since the perpetual elaboration of states of consciousness out of states of consciousness can never produce anything more than states of consciousness. But if we postulate external existence, and consider it merely postulated, the whole fabric of the argument, standing on this postulate, has no greater validity than the postulate gives it, *minus* the possible invalidity of the argument itself."

II. The next question is more vital, and lies at the root of all physical and metaphysical philosophy. Does the perceptive act, whether styled or not a state or fact of consciousness, bear witness first of all to its own existence and character, independent alike of subject and object, with a testimony no one can deny, and after this to subject and object, with a testimony that admits of denial? Such is the statement of Sir W. Hamilton, in which he professes to correct a mistake or oversight of all previous philosophers; and Mr Mill, who disagrees with him in so much, adopts it fully, singles it out for praise as a redeeming proof of his rival's great acuteness and sagacity, and says that the distinction he draws, between that part of the testimony of consciousness which is incontrovertible and controvertible, "is in the main beyond question just." I believe, on the contrary, that the doctrine is not only untrue, but, in Mr Spencer's phrase, unthinkable, that what it is said no one can possibly deny is a self-contradiction. And it is less presumptuous to hold this of a statement which Mr Mill and Sir W. Hamilton agree in making one main basis of their two philosophical creeds, in other respects so diverse, when the latter speaks of it as a discovery of his own, which has been overlooked by all previous philosophers.

To make the question at issue clear, let us take a particular case, expressed by a brief sentence. I see a house. Here we have a subject, an object, and an act of vision. In the usual and popular view, the subject and the object, the I and the house, are abiding realities. The act of vision is a momentary relation between these realities, capable of being suspended, renewed, or continued, as often as we open and shut our eyes. This act of vision Sir W. Hamilton and Mr Mill agree to call an act or state of consciousness. The former personifies it, and calls it a witness. It bears witness, he affirms, to three things. First to its own existence and character, as a testimony about a subject and object really existing, and testifying to both at the same indivisible moment of time; and next, to the existence of that subject and object. We may question whether its testimony to their existence is true. We cannot question that, whether its testimony to them is true or not, in other words, whether or not there is an I which perceives, or a house perceived, the act of vision does testify to the mind and the house as real, and to both at the same instant.

Mr Mill agrees in the fact of a present testimony of consciousness to its own reality, and that of a subject and an object. But he holds that this latter testimony is not simple and immediate, but a derived result of many past experiences. With regard to the house, and perhaps with regard to the mind also, he thinks that something has been interpolated without cause, a something useless and unintelligible, the idea of a substance or substratum of the sensations, while all that we can really know is the series of sensations or perceptive acts alone.

The true analysis of perception, I believe, differs wholly from either view. First, perception and consciousness are not the same, nor strictly simultaneous, but distinct and successive. Only one act follows the other so swiftly, that the sequence may be overlooked; and when the perceptive act is repeated or continued, which is usually the case, the two series are mingled, and melt into one.

Next, an act of vision, for its existence, requires three things, a seeing mind, an object seen, and a momentary relation between them, which depends on the bodily organization and the rays of light. But what it teaches, or the element of knowledge it supplies, is not of all three alike, but of the object alone. When I see a house, this teaches me something about the house, but not about my mind, or my eyes, or the rays of light, or the nature, mental or physical, of the momentary act of vision. The consciousness that usually follows is a reflex action. Its direct object is the mind itself, known through the act of vision, and these along with it, in their relation to each other. A knowledge of the mind's existence is the first and main part of this consciousness, and the second is the mind's memory of its own act of vision. The second, when the first is removed or denied, is impossible. If I do or did not exist, I cannot have seen a house. If I do not believe that I exist, I cannot believe that I have seen a house, or am now seeing it. An act of nothing, seeing nothing, is an impossible chimera. To sublate the person sublates every personal act. But to personify the act of vision, as a witness independent of the seeing mind, and then make it bear witness to itself first, and to the person seeing only in the second place, the former testimony being more certain and undeniable than the second, is no proof of singular acumen and sagacity, but a strange and extreme instance how able men may deceive themselves by metaphors and figures of speech. Sir W. Hamilton and Mr Mill, so often at variance, here agree. But their agreement, as the former admits, deserts the view of all previous philosophers, while it also plainly runs counter to the laws of speech, and the common and familiar belief of all mankind.

An act of vision cannot bear witness to itself. Still less can it bear witness to the sameness of the moment of time, in which I recognize the existence of a thing seen, and of my mind that sees it. The act cannot occur, or give us knowledge of any kind, unless there is a mind that sees, and a thing seen. But the preconditions of knowledge, and its contents or substance, are not the same. What is taught or seen by the act of vision is neither the mind, nor the mind's act, nor the rays of light, nor an optical image on the retina, but the exciting cause of vision, the outward visible thing alone. The reflex act of consciousness is a rapid retrospect, a self-recollection swiftly sequent on the perception. Its object is not single but twofold, as the word implies, the mind and a previous act of the mind, thought of and known together. But while we know the mind only through the memory of its act, still the knowledge of the mind holds the first place in the compound act of consciousness. I must first know that I exist, before I can know that I have seen anything. No one can possibly form any conception of an act of sight, in which nobody sees, and nothing is seen.

A perceptive act, subject and object being sublated and set aside, is a chimera and nothing more. It can be no oracle, able to announce infallible truths, beyond the power of scepticism itself to dispute. I may sublate the object alone, and put another in its place. Then there will still be the mind, a thing seen, and act of vision, only not the same as before. What I saw, and mistook for a house, may have been a rock or a tree. But if I sublate the seeing mind, every act of vision is sublated, and becomes impossible. If there is no one who sees, plainly there can be no sight of any thing whatever.

How does Sir W. Hamilton attempt to prove this prior and more certain testimony of the perceptive act to three of its own features, which must be believed, even if we doubt or reject the substance of its evidence with regard to the reality of subject and object? He says that it is no other than the consciousness itself, and to doubt the existence of consciousness is impossible. For such a doubt can exist only in and by consciousness, and so would annihilate itself. We should doubt that we doubted.

Mr Mill, who shares the doctrine, rejects this proof of it, but for a reason just as erroneous as the argument itself; namely that to doubt and not to think at all are the same thing, so that doubting is not a state of consciousness, but its negation. A very surprising statement in so able a philosopher, and one which needs no formal disproof. The negation of fixed belief or disbelief, and of all thought whatever, are clearly not the same thing. The true fault of the argument is that it confounds two senses of the word, consciousness, which are different and even opposite.

What do we mean when we say that a doubt can exist only in and by consciousness? Simply this, that a man must both exist and think, before he can doubt, since to doubt is to think in one especial way. But what is the supposed fact, called consciousness, in the other part of the argument? It is an asserted testimony of the perceptive act, first personified, to three things; its own averment of the reality of the mind and the object, the immediacy of that averment, and the strict coincidence of both its parts in time, so that neither precedes or follows. This threefold testimony, Sir W. Hamilton says, may be, and has often been, contradicted, but no one can deny that it is made, for it is the consciousness itself. I believe, on the contrary, that such an averment is not only untrue, but impossible and unthinkable. It is untrue, so far as it is supposed to affirm the immediacy of information which can be gained only through known media, the organs of vision and the rays of light. It is further untrue, so far as it maintains the strict simultaneousness of the perception, by which we recognize the object of vision, and the consciousness, by which we recognize together our own mind which sees, and its act or passive impression in seeing. A careful analysis of thought will prove that they are successive. But the averment is incredible and impossible, so far as it supposes any testimony of the perceptive act, when the acting, perceiving mind is denied and done away. The contradiction is as great as to affirm that the height of a man's stature may give infallible evidence in a court of law, even though the man, whose height it is assumed to be, does not now, and never did exist.

Mr Mill has pointed out, and fully established, two grave errors in Sir W. Hamilton's defence of Reid's doctrine and his own, of Natural Realism. The first is that he wholly misrepresents the view of the philosophers to whom he is opposed, as to the testimony of consciousness. They do not believe that consciousness declares to them and all mankind certain facts, and then take the liberty of contradicting the facts. Most of them would contradict this statement, and deny precisely what he says it is impossible to deny. They think that consciousness does not testify what he thinks it testifies. Many of them believe that consciousness testifies to nothing beyond itself; others, that it gives some testimony, but not the same he ascribes to it; others, again, that it gives the testimony he asserts, but not, as he asserts, intuitively. Thus his arguments are addressed to the wrong point, the duty of believing its witness, instead of the real question in dispute, what it is that consciousness testifies, and in what way. (Exam. p. 166.)

The other fault is more serious still, that he reverses and abandons the very doctrine of Realism he professes so zealously to defend. Nothing, he says, can be conceived more ridiculous than the opinion of philosophers, not excepting Reid, on the nature of vision. "It has been curiously held that, in looking at the sun, moon, or any other object of sight, we are either conscious of those distinct objects, or else that these objects are those really represented in the mind. Nothing can be more absurd. Through the eye we perceive nothing but the rays of light in contact with the retina."

Prof. Grote (Explor. pp. 143—5) and Mr Mill (Exam. pp. 174, 194) both express their surprise and amazement at such a statement from so earnest a champion of Natural Realism, or the direct testimony of consciousness to the existence of the outward material world. If the testimony, after all, is only to the film of air in contact with the rest of our bodies, and the rays of light on the retina or surface of the eyes, the difference from pure idealism, to common minds, is so slight as hardly to be worth a controversy. We know the circuit of our mind on one hypothesis, of our body on the other, and nothing beyond. The only proper name for such a philosophical creed is Unnatural Realism; and Sir W. Hamilton, with one stroke of the pen, annuls and reverses, as absurd, the very doctrine he professes so laboriously to defend.

The true doctrine of perception, I conceive, may be thus stated. We perceive, see, touch, hear, taste, outward material objects, the things themselves. A sensible impression is an effect, which suggests and proves at once, a real thing without us, a real cause. But the result of a single perceptive act is too vague to be called knowledge. We perceive a something, the cause of a visual or tactual effect, but what it is, or where it is, or of what kind, we cannot decide from one or two sensations alone. But when they are multiplied, since the thing, the cause of sensations, abides as well as the mind which observes, the conception of this cause becomes more and more definite. The shapeless receives a shape, the unplaced receives a known position, from the consilience of many acts of perception in various stations of the observer. Solid objects within the range of both touch and sight are first determined. And these in turn form a kind of base-line, as in a geodetic survey, whence we can pass on to define the more obscure or remote; till we learn the nature and properties of things invisible or inaccessible, as water and air, the clouds, the sun, moon and stars.

It is not the sensations themselves, which are grouped or added together. These do not co-exist. One expires before another is born. It is the outward thing, the cause at first unknown, which is identified, localized, and more and more exactly defined, by elements of knowledge derived from many sensations and separate experiences. This synthesis goes on side by side with the analysis, which separates one part or one quality of the object from another, and discerns the relations between them. Thus our knowledge of these outward things comes to be two-fold, partly of their internal relations, of one part of them to another, and partly of their relations to ourselves or other observers like ourselves, in exciting and awakening various kinds of sensations. The first are the primary, the others are the secondary properties of matter. We know the first, as in the objects, and existing apart from, and independent of, any actual sensation of touch, taste, smell, or sight. The others we know as in the body, but latent, until they come within the range of the senses of some sentient being.

In all the remarks, then, of Mr Mill on the gradual growth of our conception of a material object, out of many experiences, which afterwards cohere by a law of mental association, of which the results may remain, even when the process is forgotten, I think that his teaching is exact and sound. The one error, I believe, which vitiates the whole course of his reasoning, is that very doctrine of Relativity, which he accounts to be "true, fundamental, and of high importance." In Sir W. Hamilton its mischievous effects are partly limited by the Realism which he holds side by side with it, though the two doctrines are in direct opposition. But this again is neutralized by the second contradiction, which reduces it to a doctrine of Unnatural Realism, quite as remote from usual and popular belief as the philosophical theories he condemns. In Mr Mill, with his greater consistency of thought, the doctrine of Relativity is worked out, almost to its extreme results, in which it abolishes the world of matter, and leaves the world of mind hovering between life and death, in a state of suspended animation, neither believed nor wholly disbelieved.

The foundation of the whole doctrine is in the statement that we know directly our own sensations, and nothing but those sensations. This, I believe, is wholly untrue. A sensation is a momentary, perishable thing. It expires as soon as felt. Knowledge, properly, is not of something that ceases before we can name it. On the contrary, it can only be of something that abides and endures. We know, not the sensation itself, but something else by and through the sensation. The landscape is not the pane of glass through which we look upon it. We may look on a landscape through a thousand panes in succession, and though the panes are many, the landscape is one and the same. Each pane might be broken as soon as we have looked through it, or each opening closed in succession, but our knowledge is of the landscape, not of the panes or openings. We may look at a star through a telescope, but while the observation is made, we do not see the lenses of the telescope, but the star alone. A sensation ends as soon as felt, but it tells us of some cause without us, which exists, and does not perish in a moment. A second and a third sensation adds to our knowledge of this permanent reality. The sensations themselves cannot be grouped or tied together. They can be known only in an improper sense, as momentary, variable relations between two known things, a sentient mind and a thing perceived. And this knowledge is more difficult and complex than that of either mind or material object, and comes only by degrees in the later stages of science. The knowledge of the air and invisible gases, through which alone we see, forms not the earliest, but one of the latest stages in scientific chemistry.

The strangeness of the paradox reaches its height, when Mr Mill would replace Matter, as common minds understand it, by his new phrase, "Permanent Possibilities of Sensation." Each of the three terms, which make up the phrase, involves a separate ambiguity or self-contradiction.

The word sensation is ambiguous. It may be either active or passive. The possibility may be either of feeling or of being felt. When used as a substitute for matter, the second meaning is that required, but when sensations are used to replace mind, the first. The phrase, permanent possibilities of sensation, may indicate something permanently possible to be felt, or something permanently capable of feeling. In one sense it is a cumbrous and obscure periphrasis for matter, in the other, for mind. And thus, in its actual, abstract form, it is doubly ambiguous. It may be taken alternately to exclude or include each of the two factors, on which the possibility of any sensation depends. The non-existent cannot possibly see, and the non-existent cannot possibly be seen.

Again, possibilities imply realities. If there is nothing real, there can be nothing possible. Nay more, the very term implies and requires the notion of a power or powers not actually exercised. If there are possibilities of sensation, there must be real beings that can have feelings which do not now exist, or real things, that *can* be touched, tasted, heard, or seen, when there is no present actual sensation. Thus possibilities of sensation imply, not only the existence of minds or sentient beings that feel, but that these can feel many sensations never yet felt, and remember sensations that have ceased; and also that there are things, capable of causing many more sensations, besides those which we experience at the moment.

In the third place, permanence is a character which does not belong to sensations at all. "Present sensations," Mr Mill allows, "are generally of little importance, and are moreover fugitive." What is true of present sensations is true also of the past and the future. As sensations they do not and cannot endure. Each of them is the differential element, in point of time, of a relation between two existences, which are permanent and enduring, and still are subject to variation in their modes of being, and in their relations to other beings.

But if actual sensations are not permanent, a bare possibility of sensation is still less capable of this quality of perdurability. To avoid the recognition of matter and mind, as real existences, to which direct witness is borne in every act of perception and of recollection or consciousness, we may pile one metaphysical abstraction upon another. We turn the relation of two things into an independent reality, and try to imagine a sensation of vision, when there is no one who sees, and nothing seen. We may go further, and replace sensations themselves by mere unrealized possibilities. We may take a third step, and assign to the unreal, only possible sensations, a permanence which no actual sensations ever possess. But the only result will be to replace the natural results of human thought, embodied in every language under the sun, by a cumbrous, obscure, and perplexing phraseology, which either admits what it is sought to deny, or else involves a threefold variety of error and self-contradiction.

The doctrine of relativity, in the sense which Mr Mill despises, may be a mere truism, but in the sense which he adopts and presses it is a fundamental and mischievous error. It shuts up the mind within itself as a prison, and makes it capable of knowing its momentary sensations, while they last, but nothing more. Its view of perception is wholly deceptive and untrue. We do not feel our own act of feeling. We do not see our own act of seeing, no, nor even an image on the retina, or the rays of light in contact with our eyes, as Sir W. Hamilton so dogmatically affirms.

The senses are the windows of the soul. Through them we see the outward landscape, the realities of that wonderful world which God has created and made. We do not see the window pane or the air, still less the act of looking through them. But we see the outward objects, the material things. We see the things themselves, when we discern them as the localized and definite causes of the effects they produce on us through the senses. And we see the things *in* themselves, when we come to learn the local and causal relations of their parts to each other, as well as the properties of the whole in relation to ourselves. Our knowledge of mind is later in order of time than our knowledge of matter, but, when once reached, is the result of a larger and fuller induction. The certainty comes later, but once attained is more full and complete. And the double reality of matter and mind points upward to the Supreme Reality, the God of the spirits of all flesh, who is also the Author and Architect of the material universe. This truth is latest in order of growth, the result of a still wider induction, and complicated with more numerous mysteries. But, when once attained, it is the highest, purest, and most certain of all truths. For all lower truths depend upon it. With its help they are resolved into sacred mysteries, linked with growing insight and knowledge. When it is rejected, they sink down into a chaos of intellectual self-contradiction and moral darkness.

CHAPTER V.

ON THE REALITY OF MATTER.

PHYSICAL Fatalism, viewed as a scheme of philosophy, includes three main principles. First, that Theology must be discarded and set aside, as an unreal and impossible science. All religion, in its view, is only blind emotion, or vague yearning after the unknowable, and is not science, but pure nescience. Next, that Physics are the sole science, and material phenomena the only field of thought, in which knowledge is attainable. Thirdly, that Psychology, and all Social, Political, and Moral Philosophy are only branches of Physics. They are special applications of the general theory of matter, motion, physical force, and material or atomic change.

The first of these doctrines has now been examined and disproved. Our next inquiry must be, how far the teaching of the system, with regard to Physical Science itself, is clear, accurate, and consistent. Here, if anywhere, we may look for its strength. Its main feature is the exaltation of Physics, so as to absorb into itself all the sciences of human thought, feeling, and action. Its birth, as a philosophy, is chiefly due to the great and undoubted progress, within the last two centuries, of physical discovery. Are the foundations then, on the side of Physics, so clearly and firmly laid, as to bear the immense superstructure of positive and negative theory, which it is attempted to rear upon them?

The common sense of mankind agrees in affirming that we are real persons, living in the midst of a real, outward, and material world. The voice of Revealed Religion is the same. But it teaches further that this world of matter owes its being and its arrangement to a Divine Creator, and is subject to laws of his appointment, which man cannot supersede or destroy. That the material world really exists, and is knowable, and that much knowledge has been attained already, and still more is attainable, respecting its laws and changes, seems to be the first and indispensable basis of a Science of Physics. Here, then, we should expect the Fatalistic Philosophy to be in perfect agreement with the concurrent voice of common sense and Christian Revelation. The knowledge of matter and its laws is all that it can offer for our solace, when it has reduced morals into mechanism, and extinguished Theology. Yet here, strange to say, it hesitates and oscillates. The maxims it lays down, to exclude the possibility of religious knowledge, are like a millstone round its neck, when it sets about the task of founding a Science of Matter. Thus it seems "to say and straight unsay," to affirm and deny, alternately, the truth which is the only possible basis of Physical Science, and which once rejected, the whole range of supposed discoveries in Physics must resolve themselves into a huge pile of unreal and shadowy delusions. It speaks with stammering lips, and in a strange, ambiguous dialect, to which it is not easy to attach any clear and consistent meaning.

And first, we have these statements on the positive side, in which the real existence of matter is affirmed.

Our first belief in its objective reality, we are told (p. 93), is one which metaphysical criticisms cannot for a moment shake. "When we are taught that a piece of matter, regarded as existing externally, cannot be really known, but only certain impressions produced on us, we are yet compelled to think of these in relation to a positive cause: the notion of a real existence, which generated the impressions, becomes nascent." (p. 93.) And this belief, which persists at all times, under all circumstances, and cannot cease till consciousness ceases, is said to have the highest validity of any. (p. 94.) "The Noumenon, everywhere named as the antithesis of the Phenomenon, is throughout necessarily thought of as an actuality. It is impossible to conceive that our knowledge is of Appearances only, without conceiving a Reality of which they are appearances, for appearance without reality is unthinkable." (p. 88.) "It is impossible to get rid of the consciousness of an actuality, lying behind appearances, and hence our indestructible belief in that actuality." (p. 97.) To suppose Space subjective "is to escape from great difficulties by rushing into greater." It "multiplies irrationalities," and "is positively unthinkable in what it tacitly denies, and equally unthinkable in what it openly affirms." (pp. 49, 50.) Our conception of matter, we are told, is made up of extension and resistance; and of these the resistance is primary, the extension secondary. (p. 166.) But if Space be without us, and Space and Force together are the defining nature of matter, the inevitable inference must be the real existence of an outward, material world. And it is suggested, accordingly, that if the word *effect*, which is equally appropriate, were substituted for appearance or phenomenon, "we should be in little danger of falling into the insanities of idealism." (p. 159.)

But the negative statements are equally distinct, and perhaps more numerous. And first (p. 54), "Matter in its ultimate nature is absolutely incomprehensible. Form what suppositions we may, we have nothing but a choice of opposite absurdities." And again (p. 86), "Analysis leads to the conclusion that things in themselves cannot be known to us, and that the knowledge of them, were it possible, would be useless." In the chapter on the Data of Philosophy, the hard problem is attempted, to find a name for sense-perceptions that shall not imply the existence of anything perceived, and for states of consciousness or thought, that shall not imply the existence of any thinking, conscious mind, or for modifications of mind, without implying that there is any mind to be modified. And the answer is found in the definition, that "all things known to us are manifestations of the Unknowable."

In Part II. Ch. III. on the Knowable, further light is given on the true doctrine of the work, and the subject is thrown back into complete ambiguity and confusion. A frank admission is made of that sense of illusion which is said commonly to attend the reading of metaphysics. And the fault is said to be with metaphysicians themselves. They reject one half of a vulgar error, but not the whole. The peasant thinks that he sees a house, a tree, or a neighbour, and not a mere appearance or phenomenon, or an inward state or act of his own mind. The philosopher proves that our knowledge of the external world can be but phenomenal, that the things of which we are conscious are appearances only. Thus he transfers the appearances into consciousness, but he leaves the reality outside. He forgets that the conception of reality can be nothing more than some form of consciousness. Hence the remedy Mr Spencer proposes, to remove the popular sense of illusion in metaphysics, is a new definition of reality. "By reality we mean persistence in consciousness....Reality, as we think it, is nothing more than this persistence." (pp. 160, 161.)

Here consciousness, as with Sir W. Hamilton, Mr Mill, and others, is diverted from its proper sense, and used as a generic name for all mental thoughts, feelings, and impressions. It is simply feeling or thought. Persistence in consciousness is persistence in thought. The definition means, then, in plain English, that the only reality is constant thinking about some unknown and unknowable thing. Thought is real while it lasts, and because it lasts, even although outside the thought itself there is no real thing to be thought of.

Surely no stranger cure could be devised, than such a definition, for the complaint of simple men, that metaphysicians rob them of all reality, and leave them "floating in a world of phantasms." The insanity of idealism, as Mr Spencer has called it on the previous page, is to be remedied by reversing the meaning of words along with the instincts of common sense. There are to be no real things in the popular sense of the word. But by way of compensation, thoughts about the unreal, the unthinkable, or the non-existent, when they persist, and continue long enough, are to be counted realities by virtue of their continuance alone.

The proposed remedy, then, for the fault of metaphysics, and removing a slight blemish, merely aggravates and redoubles a great defect. We are not only to replace realities by mere phantasms, but boldly to apply the very name of realities to shadows when the substance is gone.

The sense of reality in things around us, Mr Spencer has truly said, is one which no metaphysical criticisms can shake in the least. It has grown with our growth from the first dawn of life. It results from the confluence of millions of personal experiences, confirmed and re-echoed by millions on millions of other experiences of our fellow-men. This fundamental conviction can never be reversed by a new definition of a single word. It is the conviction that our visual perceptions are the effects of some cause without us, some real thing which abides and endures. The sensations themselves are born and die with each passing moment. Their lifetime is the twinkling of an eye. But the thing seen does not perish, because we shut our eyes; nor does it begin a new life, as often as we open our eyes, and look on it once more.

"Permanent possibilities of sensation" is merely an ingenious phrase, to disguise and conceal a self-contradiction. Once let there be some actual existence, and millions of possibilities may depend upon it. But if there be no perceiving mind, and no objects to be seen, how can there be any possibilities of visual perceptions? How can it be possible for a mind to see something, when there is no mind, and no visible object? Spider's threads, though exceedingly fine, have their use in the micrometer, and may even be helpful to astronomers in learning the exact position of the stars. But no science of astronomy could exist with a universe composed only of spider's threads, stretched out on all sides, with no fastenings, in empty space. As little can there be science of any kind with a universe consisting of countless, momentary sensations, with neither starting-point nor goal, when there is no mind that sees or feels, and no object anywhere to be felt or seen.

"Reality, as we think it, is nothing more than persistence in consciousness." To think reality is a strange and obscure phrase. To think persistence in consciousness is stranger still. What seems to be intended is this. I see an orange, and hold it in my hand. There is no real thing, no actual orange, occupying a definite place, from which it excludes other solid things by its presence. In affirming this we should "transcend consciousness," and pretend to know something of the Unknowable. But I have a muscular feeling of touch, and a visual impression. Now this impression persists, so long as I keep my eyes open, and direct them in one particular way. It ceases when I close them. It persists in recurring, and is renewed, when I open them again. This pertinacious recurrence of the mingled sensations of resistance, yellowness, and roundness, is the only real thing. No real orange is seen; but there is a persistent feeling that we see a something, or have a compound sensation of hardness or softness, colour and shape, to which we give the name, orange. This definition of reality is the cure proposed in the First Principles for "the insanities of idealism." To plain minds it will seem like a climax of the disease.

Matter must exist, and be knowable, before there can be a real Science of Physics. This "first principle" of reason and common sense Mr Spencer seems alternately to affirm and deny. The positive statements which imply its existence, and a distinct conception of its nature, approach very near to a just and consistent theory. On the other hand, the doctrine of the unknowable, "like a dank mist slow creeping," settles down over the subject again and again, undoing and reversing the clearer statements, and involving the whole question in ambiguity and confusion. Consciousnesses or states of consciousness, phenomena or appearances, relative realities, absolute realities, the Absolute Reality, and the Unknowable, are named in such a way, that it is impossible to decide which are meant to be the same, and which are distinguished from each other. But if we combine and select the sounder parts of the doctrine affirmed, the others, which are either obscure and ambiguous, or directly contradict them, will drop away and disappear of themselves.

And first, we have a firm belief in the objective reality of Matter, "a belief which metaphysical criticisms cannot for a moment shake." When difficulties are started as to any special form of this conception, "though the conception be transformed, it is not destroyed: there remains the sense of reality, dissociated as far as possible from the special forms." Thus, in contrast to the view of Berkeley and Berkeley's successors, the objective reality of material substance, as something at once real, and without the mind, may be taken as the first step in a sound and true theory.

Next, our fundamental conception of matter is one of "coexistent positions that offer resistance, as contrasted with our conception of space, in which the coexistent positions offer no resistance." Thus our conception involves two elements, resistance and extension. Of these "the resistance is primary, the extension secondary." For "body is distinguished from space by its resistance, and hence this attribute must have precedence in the genesis of the idea. The resistance-attribute must be regarded as primordial, and the space-attribute as derivative. Matter is present to our consciousness (by which is meant our perceptive faculty) in terms of Force. It follows that forces, standing in certain correlations, are the whole content of our idea of Matter."

These remarks are true, but the analysis, I think, is incomplete. Extension, physically, consists of three dimensions. But logically it seems made up of two distinct elements, local position and continuity. Now local position is inseparable from our idea of matter, but continuity is a separable element. In fact, we learn to know solid bodies by their limits, in contrast with what seems empty space around them. In the first stage of our experience, body, which we can both see and touch, is contrasted with empty space, which can neither be touched nor seen; and air and empty space are equivalent. In a further stage, the same relation of distance, which separates bodies from ourselves and from each other, is seen to exist between the parts of the same body. The analysis is then carried still further. We learn to recognize the existence of pores, or parts of the collective bulk of solid bodies, which are not body, but parts of space or intervals merely, and also that aerial space, which is empty to the eye, is partly occupied with matter.

When this process of thought reaches its limit, we attain the conception that extension is a relation of distance, either between whole bodies, or parts of the same compound body, and does not apply to the constituent units or atoms, but to composite structures alone. On the other hand, position and force remain as defining and constituent elements of the units themselves. And thus our definition of matter becomes modified. Distance is recognized as a relation between its parts, and a main constituent in the true conception of a material structure or compound body, but not of the ultimate units or atoms. Place, however, remains as a necessary element. And it is not secondary, but primary. We can conceive of coexistent positions without any force; and in fact, the whole science of Geometry is a development of their relations and properties. But a science of Force, from which the conceptions of geometry are excluded, is impossible. Thus it is not strictly true that the resistance-attribute is primary and the extension-attribute secondary, in a just conception of matter. Extension has two elements, position, and distance or continuity. Position is a more essential and primary element than force itself, because we can conceive positions without force, but not forces without direction or position. On the other hand, distance, after being first recognized between ourselves and solid objects, and those objects themselves, is next recognized between the sensible parts of bodies, then between the insensible, till we reach the final conception, that it is a relation between the material units, and not an inherent attribute of those units themselves.

Matter, then, in the final analysis, is a sum or integral of localized forces, or of dynamized positions, points that are force-centres. And the second form of the definition is the more correct, because we can conceive of positions and dimensions without force, but not of forces without local direction, nor of central forces without a centre to which they belong.

Place, then, being one of two essential elements in our conception of matter, its outness and reality, according to other statements in the First Principles, will be clearly established, "The direct testimony of consciousness," we there read, "is that space is not within, but without the mind," and an opposite view "is unthinkable in what it tacitly denies, and equally unthinkable in what it openly affirms, is impossible to realize in thought, and multiplies irrationalities." Since Space, then, is external, and cannot be otherwise thought of, and Place enters into the very definition of Matter, it follows that Matter is both real and external.

This conclusion, however, reveals the error and entire misconception involved in many other statements of the First Principles on the very same subject.

And first, Matter is not unknowable. Its nature is defined by the union of two elements, each definite, and capable of being the object of definite knowledge, position and force. So far as it consists of units, it is the subject of the science of number, or Arithmetic. So far as these units are local, or points having local relations of distance to each other, it is the subject of a second science, Geometry, and this science, as is well known, like the first, has an almost infinite range. So far as it relates to forces, it is the subject of a third science, Dynamics, not independent of the first or second, but dependent on them both, and yet including a vast and spacious building of additional truths.

Matter, then, plainly belongs to the Knowable, not the Unknowable. Only this knowledge is partial, and not exhaustive or complete. We know it in part only. We can compare the lengths of lines or intervals of distance, or the sizes of figures and solids, only by the help of numbers. But number is infinite. And the simplest relations in geometry, such as the diagonal of a square, compared with its side, the circumference of a circle, compared with its diameter, cannot be expressed by finite numbers, but by a decimal of infinite terms. And hence we find that the true harmony of the three sciences to which Matter gives rise, cannot be found within finite limits. It exceeds the range of every finite understanding, and belongs to the Infinite Wisdom alone.

A second statement, disproved by the previous definition of Matter, is that "a centre of force without extension is unthinkable." The exact reverse, I believe, on closer thought, will be found to be true. An extended centre of force is a contradiction in terms. Again, Mr Spencer asserts that the only law conceivable for force is that it shall vary as the inverse square of distance. A very strange misstatement, considering how large a part of the Principia and other works on Dynamics, is spent in reasoning out the consequences of many other laws of force. But forces varying by this law, or by any other power of the distance, are inconceivable, except with a point for the centre of force. For from one bulk or mass to any other the distance is not one only, but the number of distances is infinite.

All sensible objects are compound and not simple. Extension, that is, distance of one part from another, is one constituent and essential element in the conception of them. But with the units or atoms it is no part of their essence or definition. It is simply a relation they bear to each other.

Again, it follows from the previous conclusions that Matter is no sum or integral of successive consciousnesses. Consciousness is of the mind only, and its own states or phases. But different places or positions co-exist, and are not successive. If the idea of space being within the mind, or a form of thought, is unthinkable, whatever is defined by the union of Place and Force cannot be within the mind. It cannot be one or many "states of consciousness," but must be something external.

Again, Matter cannot be merely a set of phenomena or appearances. Philosophy is said to prove that our knowledge of the external world can be but phenomenal, that the things of which we are conscious are only appearances. It is from this correction of the error of the peasant, who thinks that he sees the things themselves, that the notion of the illusiveness of metaphysics is said to arise. Good pictures shew us that the aspects of things may be nearly simulated by colours on canvas. So the sense of illusion is to be removed by altering the very definition of reality, and pronouncing it to be nothing more than persistence in consciousness. This is equivalent to an assertion that our often thinking of something unreal and non-existent is the only reality.

Apart from this great paradox, appearances ought to be distinguished both from states of consciousness and from realities. They consist of some part or parts of our complex notion of a visible thing, abstracted from the rest. A picture is an arrangement of colours on a flat surface, designed to agree, in reference to an observer, with the angular or perspective relations of some real figure, scene, or landscape. The sense of sight, alone, does not make known the distance, but colour and angular relations alone. On this evident truth the science of perspective is founded. Hence appearances may be real things, viewed in reference to their surfaces only, our knowledge of their internal structure being gained in some other way; or false apprehensions of real things, referred to a place which is not their true place, as in the reflections of a mirror; or such as suggest and represent an object not truly present, as in the case of pictorial representations.

Once more, the Matter we know, whether we style it a relative or an absolute reality, is not the same with the Mind or Self which we also know. And neither of them is the same with the Absolute and Infinite Being, who is the object alike of the highest philosophy, and of all religious faith. The attempt, in the chapter on the Data of Philosophy, to organize states of mental experience, and classify them in two main contrasted sets or streams, while claiming to dispense with either of the two main postulates, a conscious mind of which they are states, or real objects of which they are the perceptions or appearances, can be nothing else than a series of logical illusions. The new title, "manifestations of the unknowable," is invented in order to escape from the need of assuming the existence of real Mind, or real Matter. These latter postulates, it is owned, are accepted by mankind at large with a belief that is irresistible. And the postulates offered to replace them are these: that we know and are sure of the existence of something wholly unknowable; that this Unknowable may be manifested or made known; that all phenomena are manifestations of one and the same Being, of which we cannot know whether it be many or one; and that self and not-self, the ego and the non-ego, are one and the same thing. But these new postulates are self-contradictions, pseud-ideas, and unthinkable in the most true and proper sense of the term.

Sense-perceptions, in the Data of Philosophy, are surnamed "vivid manifestations of the Unknowable." In this description the second and third terms neutralize and destroy each other. And the epithet, vivid, belongs to them, only from the force with which they suggest to us the actual presence of a real object. Again, what are called "the faint manifestations," that is, reflections or internal thoughts, have for their first ground and defining element the self-consciousness of a mind that thinks and feels, whose acts, reflections, and feelings they are. It is just as easy to devise a science of plus and minus signs, entirely excluding from our thoughts any numbers or quantities to be added or subtracted, as to frame a science of sense-perceptions and reflections, by whatever name we please to call them, when we have first sublated and removed all real objects that may be felt, seen, or handled, and any real mind to which the internal meditations and reflections belong.

"Appearance without reality is unthinkable." This is true, when once the conception of distance has been gained by actual experience. We are then constrained to localize, more or less exactly, the causes of all our visual impressions. When we refer them to the true distance, as well as the right direction, we have a genuine act of vision. We see an object, and see it where it is. When we refer the object to a wrong distance, or place it in thought in a wrong direction, we have a visual illusion. We may refer the object to a right distance, but a wholly wrong direction. This is the case of images, seen by reflection in a mirror. Or again, we may see them in a right direction, but refer them illusively to a wrong distance, or set of distances. This is the case with pictures, in which the laws of perspective are observed. But there can be no appearance without the suggestion of some real object. In the case of illusive appearances, something exceptional makes the common law of association untrustworthy, and they require to be corrected and revised by a new set of experiences. We may thus learn to refer the images in a mirror correctly to their prototypes, and to realize together the perfect representation of a building in a photograph, and the fact that it is a representation, and nothing more.

Matter, then, is neither a series of consciousnesses or mental acts, nor of appearances without reality. But we are told that it is a relative, and not an absolute reality, at least the Matter that we know. It is not easy to reconcile this with the new definition proposed. What can be the meaning of a relative persistence in consciousness, and an absolute persistence in consciousness? A consciousness must be relative, for it relates to a conscious mind. But what is an absolute persistence in consciousness? The natural meaning of an absolute existence is that something exists, whether we are thinking of it or no. Let us hear Mr Spencer's explanation of his own very obscure phrase.

"Such being our cognition of the relative reality, what are we to say of the absolute reality?......It is some mode of the Unknowable, related to the Matter we know as cause to effect. The relativity of our cognition is shewn alike by the analysis, and by the contradictions that are involved, when we deal with the cognition as an absolute one......Though known to us only under relation, Matter is as real, in the true sense of the word, as it would be, could we know it out of relation. And further, the relative reality we know as Matter is necessarily represented to the mind as standing in a persistent or real relation to the absolute reality. We may therefore deliver ourselves over without hesitation to those terms of thought which experience has organized in us. We need not in our researches refrain from dealing with Matter as made up of extended and resistant atoms; for this conception, necessarily resulting from our experience of matter, is not less legitimate than the conception of aggregate masses as extended and resistant. The atomic hypothesis as well as the kindred hypothesis of an all-pervading ether, is simply a necessary development of those universal forms, which the actions of the Unknowable have wrought in us. The conclusions, logically worked out by the aid of these hypotheses, are sure to be in harmony with all others which these same forms present to us, and will have a relative truth that is equally complete."

This passage is an enigma, which needs an Œdipus to explain it. It seems to me sown with incongruities and contradictions, as thick as the stars in the Milky Way. And first, how can we know a relative persistence in consciousness, if that is the only meaning of reality, and how can the knowledge of a relative persistence be diverse from that of an absolute persistence? If we go on thinking at all, our thought, as a fact, is absolutely real, and yet it is plainly relative to a thinking mind, even if we are the victims of a mere illusion, and there is no real object whatever, of which to think assiduously. The absolute reality is "some mode of the Unknowable, related to the Matter we know, as cause to effect." Here many questions arise. Can we know that the Unknowable has many modes? Can the unknowable matter, stand in relation to known matter, as cause to effect, and still remain unknown? Can the mind see double, so that for every object around us there is one known thing, and another distinct from it, known to be its cause, but still wholly unknowable? For instance, I see and touch a book or an orange. Is there a known and knowable book, which I see and touch, and another book, unknown and unknowable, which I neither see nor touch, but which is the cause of the first, and attends it everywhere like a shadow? "Though known to us only under relation, Matter is as real to us in the true sense of that word, as it would be, if it could be known to us out of relation." Knowledge, of course, is a relation between him who knows and the thing known. Matter then, may well be real to us in the true sense of the word, though known in relation, for knowledge itself is a relation, and there must be this relation, if there be real knowledge. But then what can be meant by the alternative suggested? The real knowledge we have is as real as the imaginary knowledge, by which we may be conceived to know a thing in itself out of relation, that is, to know a thing without our knowing it, and without the thing being known. "The relative reality we know as Matter is necessarily represented as in a real relation to the absolute reality." Now the matter we know has been just defined to be a sum of correlated forces, localized, or fixed to a definite place, and the limits of some definite figure. This, in fact, is the matter of common sense, and of all mankind. It is not a series of phenomena, but a localized cause of sensations and perceptions, the parts of which have also definite relations to each other. But the cause of this cause is not matter at all, but the First Cause only. The persistent relation is not one of resemblance, but of causal dependence, of the thing on some thinking mind, of the creature locally limited, and into whose being this limitation enters as one part of its definition, on the supreme and unlimited Fountain and Source of all created being.

Again, the conception of matter, as made up of extended and resistant atoms, is said to result necessarily from our experiences of matter, and therefore to be legitimate. Yet we have been told before that this very conception is inconceivable, and one of three equal, but inconsistent absurdities, and that there is no reason for choice between this and two other hypotheses, equally but not more absurd. How can we have experiences, then, of what we have never seen, of the invisible atoms of unknowable matter? How can it be lawful for us to accept an absurdity, and to reason upon it as the proper basis for a scheme of Physical Science?

Sceptical criticism, it is said finally, brings us round to a transfigured realism. "Getting rid of all complications, and contemplating pure force, we are compelled vaguely to conceive some unknown force as the correlative of the known force. *Noumenon* and *phenomenon* are here presented in their primordial relation, as two sides of the same change, of which we are obliged to regard the last as no less real than the first."

Sound criticism does certainly bring us round to realism at the last. And this realism is in threefold harmony with the instinctive belief of all mankind, with the laws of speech in every known language, and with the direct voice of revelation. It tells us that we are placed in the midst of a real world, of countless material objects, both in the earth around, and in the skies above; that our own mind is real, and that there are other minds or persons, those of our fellowmen, who can see, touch, taste, observe, and reflect on their own thoughts, like ourselves. Natural Religion adds dimly, and Revelation more plainly, a third and higher truth, that there is a real, self-existent Being, the I AM, the Supreme Reality, who is the God of the spirits of all flesh, and the Creator of heaven and earth. It teaches that He cannot be exhaustively known, because "His wisdom is unsearchable, His greatness beyond our capacity and reach;" but also that He can be effectually known, that such a knowledge of Him is our duty and really attainable, that ignorance of Him is guilt, and that to know Him truly is life eternal.

In contrast with this simple Realism, every form of philosophical Non-realism, whether of Berkeley, who sets aside the reality of Matter, or of Hume, who sets aside Mind also, or of Kant, who makes Space a form of thought or mental affection, or of Mill, who adds to Berkeley's view a new phrase, and replaces Matter by permanent possibilities of sensation; as well as the Unnatural Realism of Hamilton, who contends vehemently that we see material objects themselves, directly and intuitively, and then affirms that what we see is only rays of light in contact with our eyes;—these all will be found as groundless in philosophy, as they are opposed to the common sense of mankind.

Noumenon and *phenomenon* are ambiguous terms. Both express realities, when we use them to denote some real thing that we inwardly reflect upon, and some real thing that appears to us, and is apprehended by our senses, but with an incipient and imperfect knowledge. Both are unreal and untrue, if we denote by them a substance existing alone, devoid of all attributes or relations to other things, and attributes existing apart from any substance or actual being to which they belong. Sensations expire with the passing moment. They can never, even with the help of the most subtle metaphysicians, set up in trade on their own account, and form a universe in which there is no God, no mind, and no matter, but an endless phantasmagoria of feelings and possibilities of feeling, with nothing to be felt, and no one to feel. We must get rid of all these complications of an erring philosophy, this floating chaos of mist and phantasms, and return to the Natural Realism, which all men have been learning from their first hours of childhood, and can never unlearn, before a science of Physics can be really founded. Its first principle is that we are real persons, living amidst a real world of material objects distinct from ourselves. And this double truth leads upward to One who is the cause both of matter and mind, the Supreme Reality, who dwells in light inaccessible, but who can reveal Himself, and has revealed Himself, in love and mercy to the souls He has made.

CHAPTER VI.

THE INDESTRUCTIBILITY OF MATTER.

The negative maxims of modern Fatalistic Philosophy have now been sifted, and proved wholly fallacious and untrue. Its doctrine of the Unknowable, borrowed from other sources, but used to clear a site for the new temple of Physical Science by sweeping away every form and vestige of religious faith, has been found to involve Physics and Theology in one common ruin. Its illusive metaphysics must be dealt with, as they propose to deal with all revealed and natural religion, and cleared out of the way, by carefully distinguishing mystery from falsehood and self-contradiction, before any steps can be taken towards constituting a genuine Science either of Matter or of Mind.

The first step in this palinode has now been taken. What is done obscurely and ambiguously in the three first chapters of the Doctrine of the Knowable, has been placed, I hope, in clearer relief. Matter really exists, and can really be known. From this truth, which the work begins by denying, and then admits by stealth under a veil of ambiguous terms, we may set out anew. In this principle genuine Physics and all sound Theology both agree, the Word of God, and the common sense of mankind.

The positive maxims of Physical Fatalism have next to be examined, or its Doctrine of the Knowable. Its main pillars are these, the Indestructibility of Matter, the Continuity of Motion, the Persistence of Force, the Resolvability of Matter and Motion into Force, the Transformation and Equivalence of Forces, the Equivalence of Physical Force with Consciousness, Thought, and Will, and lastly, the fixed, determinate, and fatal character of all material and mental change.

First, Matter is indestructible. This maxim, it is said, which would once have been counted a self-evident error, has now come, by the progress of science, to be an axiom of philosophy, and not only an inductive and *à posteriori*, but even an *à priori* and necessary truth. The very highest claims are advanced on its behalf. It is not only true, certainly true, but the opposite is inconceivable. The name of thinkers is refused to the philosophical heretics, who decline to adopt it for their creed. "Definite conclusions," it is said, "can be reached only by the use of well-defined terms. Questions concerning the validity of any part of our knowledge cannot be profitably discussed, unless the words *knowledge* and *thinking* have specific interpretations. We must not include under them whatever confused processes of consciousness the popular speech applies them to, but only the distinct processes of consciousness. And if this obliges us to reject a large part of human thinking as not thinking at all, but merely pseudo-thinking, there is no help for it."

It is a severe law of speech to lay down, that wrong reasoning is not reasoning, and that men do not think at all, when their processes of thought are confused, and they think amiss. This rigid rule does not sound over-modest, when the parties to whom it is applied include Bacon and Newton, the whole Church of Christ, and the great majority of mankind in every former age. I shall submit, however, to this new law, and proceed to shew that, in Mr Spencer's remarks on the Indestructibility of Matter, the processes of thought are thoroughly confused, and that, by his own rule, there is no genuine thought, but only pseudo-thinking from first to last.

And first, what are we to understand by this doctrine? There are three things, in which nearly all men, peasants or philosophers, sceptics or Christians, will agree. Matter in the concrete, that is, in those specialities of arrangement which names define and our senses recognize, is destroyed continually. A sheet of paper is burnt, a heap of gunpowder is fired, a drop of water is evaporated, a cloud is dispersed by the sunbeams; and the sheet of paper, the gunpowder, the drop, the cloud, cease to exist. The matter may survive in other forms, but the form from which its name was derived is destroyed and is no more.

Next, the substance, in these destructions of the form which determines our sensations, does not pass away, and is not destroyed, at least within the limits of our experience. Observation and experiment reveal constantly the survival of the matter itself, as marked by weight, in other forms. Still further, what is true within the limits of our known experience we infer to be true beyond those limits. It is natural to assume that what is true as far as our observations extend is true still further, and to the utmost bounds of the universe, till some presumptions can be raised for an opposite view.

But the doctrine now examined goes much further.

It is that Matter is self-existent and eternal by a necessity of human thought; that the idea of its creation or annihilation, or of any part of it, is pseudo-thinking, and not real thought, being an essential contradiction. It is, in short, that God himself, if we may venture so far to know the Unknowable as to assume that there is a God, cannot create and cannot destroy a single speck out of the countless atoms of the universe. That such is the doctrine affirmed will be plain from the following extracts.

"If we analyse early superstition,...we find one of its postulates to be that by some potent spell Matter can be called out of nonentity, and be made non-existent. If men did not believe this in the strictest sense, they still believed that they believed it. Nor indeed have dark ages and inferior minds alone betrayed this belief. The current theology, in its teachings respecting the beginning and the end of the world, is clearly pervaded by it; and it may even be questioned whether Shakespeare, in his poetical anticipation of a time when all things shall disappear, and 'leave not a wrack behind,' was not under this influence." "The annihilation of Matter is unthinkable for the same reason that the creation of Matter is unthinkable, and its indestructibility is thus an *à priori* cognition of the highest order." "As before we found the commonly asserted doctrine, that Matter was created out of nothing, to have been never really conceived at all, so here we find the annihilation of Matter to have been conceived only symbolically, and the symbolic conception mistaken for a real one."

Such, then, is the indestructibility of Matter affirmed, and taken for the first datum of the new philosophy. It is not the fact, admitted by all, that material substance, as tested by weight, is never known to perish when its form is changed and some visible object disappears. It is that creation or annihilation of the substance of Matter is impossible in its own nature, cannot even be thought of; so that, if God exists, it is impossible to God as well as man.

On what ground is this assertion made to rest? On two or three sentences of a former chapter, where self-existence and creation out of nothing are both affirmed to be "unthinkable."

Now what is the main drift of that previous argument? It is to prove, first, that "the Power which the universe manifests to us is utterly inscrutable;" and next, that matter is just as incomprehensible as this Supreme Power, and that "frame what suppositions we please" about its nature, we have nothing open to us but "a choice between opposite absurdities." And what now is the inference drawn from these two premises? It is this:—that a Being of whose nature, and the extent or limits of whose power, we are totally ignorant, cannot create or destroy a single atom of matter, a substance of which also we are wholly ignorant, of which we cannot tell whether it has atoms or not, or think of it in any way without absurdity and contradiction. We do not know of God who or what he is. We do not know of his power whether it has limits or not, or if it has any limits, what they are. We do not know of real matter what it is, and can form no consistent idea of its substance. And yet we are told that it is "an *à priori* cognition of the highest order" that this utterly inscrutable Being has not made, and cannot either make or destroy, a single particle of this utterly inscrutable substance!

This is pseudo-thinking of the highest order. It offers us, not a choice, but a concentration of absurdities. The whole Christian Church, and nearly all mankind of former ages, are condemned for something like idiocy, in pretending to believe, and even accepting as the first article of their creed, a plain impossibility. The argument by which they are convicted of this great folly consists of two premises and a conclusion. Each of the premises separately is untrue, and the first is not only untrue, but a plain self-contradiction. And again, if both premises were true, the alleged conclusion would not follow from them, but one diametrically opposite. The proper and logical inference would be, that we cannot possibly tell whether this unknown something we call Matter was or was not created out of nothing by this unknown Power, which Christians call God, and whether or not by the same unknown Power this unknown something may be reduced to nothing again.

Again, the doctrine laid down is that Matter is twofold, an absolute reality, of which nothing can be known, and a relative reality, which may be known. These two, indeed, are said to be related to each other, as the cause to the effect, and hence our total ignorance of the first, though asserted in words, is practically denied. To know that it is the cause of certain definite effects is clearly a partial knowledge. If those effects are numerous, this partial knowledge may be very extensive. But in affirming that Matter cannot be destroyed, and that this is an *à priori* cognition of the highest order, the reference must surely be to matter as knowable, and not to the absolute reality which is pronounced unknowable. To make any definite assertion about the last, as either destructible or indestructible, is to contradict and abandon the whole theory.

What, then, is the relative reality, or knowable matter? We have this distinct answer: "Philosophy proves that our knowledge of the external world can be but phenomenal, that the things of which we are conscious are appearances." It is this "phenomenal existence which we can alone know." (pp. 158, 9.) The metaphysician "is convinced that consciousness cannot embrace the reality, but only the appearance of it." And if he is in danger of still thinking of some reality as left outside, Mr Spencer hastens to correct the mistake, and defines reality to be persistent consciousness. A something of which we can know nothing may lie beyond. But within the sphere of knowledge there is nothing but appearances, sensations, or states of consciousness. These are real things, when they persist, or continue to appear. They are unreal or non-existent, when they cease to appear, and because they so cease; for the only true definition of reality is continued or persistent appearance.

Now let us turn to the question before us. Is Matter, as knowable, destructible or indestructible? The plain reply, on the principles of the sensational philosophy, is, that it can be destroyed and is destroyed continually. The set of sensations from a leaf of paper cease when it is burnt. Those which define a drop of water cease when it evaporates. Our sensations from a cloud cease, when it "melts into the infinite azure" of the clear blue sky. Of Matter, as defined by the theory, the only matter of which it permits us to know anything, the destruction is so far from being inconceivable, that it is an evident fact of daily and hourly occurrence.

Mr Spencer writes on this subject as follows:

"The doctrine that Matter is indestructible has become a commonplace. Whatever may be true of it absolutely, we have learnt that relatively to our consciousness, Matter never comes into existence or ceases to exist.

Cases which once gave an apparent support to the illusion that something could come out of nothing, a wider knowledge has one by one cancelled. The comet that is all at once discovered in the heavens, and nightly waxes larger, is proved not to be a newly created body, but one that was till lately beyond the range of vision. The cloud which in a few minutes forms in the sky consists of substance that previously existed in a more diffused and transparent form. Conversely, seeming annihilations of matter turn out, on closer observation, to be only changes of state."

Here, then, we are brought to this plain contradiction. Matter, as knowable, is declared to be not the unseen reality, but the sensible appearances, or phenomenal Matter alone. Phenomenal Matter, it appears from daily and hourly experience, appears and disappears, perishes and is new-created continually. And yet we are told that the indestructibility of Matter has become one of the commonplaces of science. The cases of seeming destruction resolve themselves into proofs that the same matter still exists, only in an altered form. But Matter, as phenomenal, does disappear and is destroyed. The sets of sensation which make up our conception of the cloud or the drop, pass away. The cloud vanishes, the star sets, or a mist blots it out, the drop evaporates, the ship melts into the yeast of waves, the candle is burnt away and comes to an end. The substance may last in another form, but the phenomenon or appearance is gone. According to the doctrine under review, we can know nothing of the substance, the absolute reality. We must thus remain quite ignorant whether it be destructible or indestructible. Of phenomenal Matter, that is, of the appearances, we may learn something. They come within the sphere of the knowable. We are asked to accept it, as a cognition of the highest order, that this knowable Matter is indestructible, that to place a limit to its continuance is impossible and incredible. But the plain fact is just the reverse, for this phenomenal matter perishes and is renewed daily before our eyes. Thus, by the theory, of Matter, the Noumenon, we know nothing, and therefore cannot know that it is indestructible. Of Matter, the Phenomenon, we may know much. And one main thing we know of it, proved by hourly experience, is that it both may be and continually is destroyed. For an appearance is destroyed and perishes, when it ceases to appear.

What, then, is the true nature of that discovery which is said to have resulted from the growing researches of science? "It has grown into an axiom of science, that whatever metamorphoses matter undergoes, its quantity is fixed." What is the real key to the progress of this idea, and its proper meaning? It means that an entire reversal of the doctrine in the First Principles is the first step to a right understanding of modern physical discoveries. If we can know appearances only, and what lies behind them is unknowable, the earlier notion of Matter, as often perishing, and often coming into new being, is right, and the opposite notion, that it is indestructible, is wholly wrong, being disproved by countless facts of daily experience. On the other hand, the permanence of Matter, the truth revealed by science, depends on these four axioms: that Matter is not phenomenal, but the cause on which phenomena depend; that while phenomena vary from moment to moment, the cause abides and endures; that this cause is knowable, and consists of position and force joined in one; and that while the sensible effects which result from the coherent relations of its atoms to each other vary immensely, causing appearances, disappearances, and reappearances, the total amount of Matter, as tested by weight, remains unaltered. In short, Noumenon Matter, though not indestructible, is permanent, and indestructible by man. But while this is a truth, known *à posteriori*, by a long and ever-growing induction, the theory is doubly false which calls it an *à priori* truth, and affirms also that the Matter of which it is true is wholly unknowable. Phenomenal Matter is the only matter capable, by the theory, of being known. And this is not indestructible, nor even permanent. Its emblems are the soap-bubble, the dew-drop, or the cloud. The bubble bursts, the dew-drop exhales, the cloud melts into the blue azure, and vanishes away.

The first step, then, of advancing Physics must be over the grave of this Doctrine of the Unknowable. The change from the early and more crude conception of Matter, as something which perishes hourly, and is hourly created anew, to the riper view of modern science, is simply a progress from the phenomenal to the real, from momentary effects to permanent causes, from placing the essence of matter in phenomena or visual appearances to recognizing it in units or atoms that abide, and are separately invisible, objects not of sense at all, but of the analytic reason. It is in the region of the Noumenon, and not the Phenomenon, of the falsely called Unknowable, and not of the falsely called Sole Knowable, of things and not sensations, of atoms and not surfaces, of localized forces and not of outward appearances, in that very region which the theory hands over to nescience and eternal darkness, that the chief discoveries of modern Physics have their native home. The progress of astronomy was halting and slow, so long as the view was confined to the phenomena, or to the simple registering of the heavenly motions. It was when Newton passed from contemplation of the motions to that of the forces by which they are caused, and the laws of *their* variation, that the greatest step of advance was made in the progress of Physical Science, which had ever occurred from the beginning of time.

Three reasons are further alleged, to prove that the indestructibility of matter is an *à priori* truth, of which the negation is unthinkable. The first is the accumulation of experiences, the second, the incompressibility of matter, and the third is the very nature of thought. Besides this threefold argument, we have an attempted explanation how it is that nearly all mankind came into the strange position of believing that they thought what is really unthinkable, and had firm faith in something incapable of being really believed; and finally, a resolution of the doctrine into another, the indestructibility of Force. Let us examine each topic and argument in succession.

I. First, "the indestructibility of Matter is now recognized by many as a truth of which the negation is inconceivable. Habitual experiences being no longer met by counter-experiences, as they once seemed to be, but these counter-experiences furnishing new proof that Matter exists permanently, even where the senses fail to detect it; it has grown into an axiom of science that, whatever metamorphoses Matter undergoes, its quantity is fixed. The chemist, the physicist and the physiologist not only one and all take this for granted, but would severally profess themselves unable to realize any supposition to the contrary....It is proved experimentally to be an absolute uniformity within the range of our experience. But absolute uniformities of experience generate absolute uniformities of thought. Does it not follow, then, that this ultimate truth must be a cognition involved in our mental organization? An affirmative answer is unavoidable."

Here, then, we have this startling principle laid down. Whatever we have experienced many times, and never the opposite, the mind is compelled to accept as a necessary truth, and to hold that an opposite experience is not only unlikely, but impossible and inconceivable. At what stage, we may ask, in the process of observation does this wonderful transformation occur? It can scarcely be held to take place after ten or twenty experiences. If something has occurred twenty times, it can hardly be laid down as an *à priori* truth, that it must always happen, and that a contrary experience is inconceivable. Where, then, must we place the limit? Is it after a hundred or a thousand, after ten thousand or a million experiences, that the alternative, plainly conceived in thought, so many times, up to that moment, but not realized as a fact, suddenly becomes inconceivable, and a self-contradiction?

The idea is really preposterous. The very facts Mr Spencer adduces to establish the *à priori* nature of the doctrine are a complete and effectual proof of the reverse. They shew that it is simply an induction from a very large, but limited experience. And they shew further that, in every step of that wide induction, the destructibility has been conceived, and has even, in the first stage of the inquiry, seemed more probable than an opposite view. They prove further that, if Mr Spencer's main doctrine is true, and the phenomenal alone is knowable, Matter is so far from being indestructible that its constant destruction and reproduction would be one of the most patent facts of hourly experience. There may perhaps have been some chemists, physicists, or physiologists, who have inferred the *à priori* nature of the doctrine from the facts which distinctly and emphatically disprove it. But to impute this extreme folly to the whole body is a groundless and unwarrantable calumny.

But "absolute uniformities of experience generate absolute uniformities of thought." Aphorisms of this kind have a plausible sound, but, as Mr Mill says on another subject, are empty of the smallest substance. Uniformities of experience *are* uniformities of thought. The thoughts which they generate must be inferences, and inferences may be true or false, right or wrong. Of which kind is the inference here drawn? I see, let me say, the Senate-house. I have seen it a hundred, or a thousand, or ten thousand times. I saw it both yesterday and to-day, and I have seen it forty years ago. My experience is absolutely uniform. I see it whenever my eyes are open and I pass that way. Can I safely infer that it has been on that spot from all eternity, and will continue to be there world without end? Am I bound to conclude, because I have seen it so often, that its disappearance or destruction is unthinkable? Such is the thought which my experience must generate, if Mr Spencer's phrase has any real meaning. But no inference could be more plainly erroneous. We must pass from the mere repetition of a fact to the detection of a law before experiences can ripen into a science; and that science itself cannot rise above the conditions of its birth. The law, which unifies the experience, cannot possibly claim an *à priori* character, since it rests on experimental induction alone.

II. The Incompressibility of Matter is made a further reason for affirming its indestructibility to be an *à priori* truth, or a "cognition involved in our mental organization," in these words:

"What is termed the ultimate incompressibility of Matter is an admitted law of thought. Though it is possible to imagine a piece of matter compressed without limit, however small the bulk to which we conceive it reduced, it is impossible to conceive it reduced to nothing. While we can represent the parts as indefinitely approximated, and the space occupied as indefinitely decreased, we cannot represent to ourselves the quantity of matter as made less. To do this would imply that some of the constituent parts were in thought compressed into nothing, which is no more possible than the compression of the whole into nothing. Whence it is an obvious corollary that the total quantity of matter in the universe cannot be conceived as diminished, any more than it can be conceived as increased."

In this argument the premise, the conclusion, and the corollary are the same, namely, that matter, in whole or part, cannot by compression be reduced to nothing. It is assumed of the whole, to shew that it is equally impossible to conceive it of a part; and then this proof with reference to a part, conversely proves it of the whole. And the corollary infers it of that greater whole, the Universe. It is assumption from first to last.

Four axioms are affirmed, by means of which the Indestructibility of Matter is to be promoted from a result of induction into a necessary, *à priori* truth. They are these; that we can form a definite conception of the quantity of matter, and its fixed amount; that we cannot conceive this quantity varied, and either increased or diminished; that we can conceive its indefinite compression without limit; and that we cannot conceive it so compressed as to cease to exist. These axioms all contradict the doctrine of the Unknowable previously laid down, and they also contradict each other. The test, also, here proposed, by which to try whether extinction of matter is conceivable, is an exact reversal of the only reasonable test.

First, we are plainly dealing here, not with phenomena, which confessedly expire as soon as they are born, but with matter, the substance, the object of thought as a reality, which abides and endures. Of this substance we have been assured that it is wholly incognizable. We cannot think of it, without accepting one or other of "opposite absurdities." Yet now we are taught that we can have a clear and definite conception of its quantity, that we can know that compression leaves this quantity unchanged; that it is of unlimited compressibility; and that compression, though carried to its furthest limit, may annihilate the bulk, but not the substance. This is surely a large instalment of that self-contradictory article, the knowledge of the Unknowable.

Again, it is here affirmed that we can conceive the compression of a material object without limit. On the Newtonian hypothesis this is untrue. Suppose a bulk to consist of one-tenth part of solid atoms, and nine-tenths of pores or vacuum, the density might be increased nearly ten times, till all the solid atoms touch, and hardly any void is left between them, but no further. For the statement to be true, we must accept the notion of Boscovich, that matter consists of unextended centres of force.

Now this is the hypothesis which Mr Spencer has rejected shortly before in the most decisive way. It posits, he says, a proposition which cannot be represented in thought. It is unthinkable, a symbolic conception of the illegitimate kind. To suppose it is utterly beyond human power. But it is not beyond his own power, just a hundred and twenty pages later, to make it one of four axioms on which he bases his demonstration, that the indestructibility of matter is no mere result of induction, but an essential law of human thought, a cognition involved in our mental organization, and a grand *à priori* truth.

These axioms, too, contradict each other. If matter can be compressed without limit, it can be compressed until its bulk is nothing, which is one form of its annihilation. No finite force, it is true, might be adequate. But if an infinite force can crush the whole universe into a mathematical point, this would be, to all practical intents and purposes, its annihilation.

Again, the doctrine is that the quantity of matter cannot be increased or diminished. But this assumes some measure of quantity to exist, and be capable of being applied. It is presently argued that no natural unit exists. The practical approach to it is weight. Where the weights are equal, we assume the quantities of matter to be equal. But in the light of science this test fails. Weight varies, though the material atoms are the same. A ball of iron will lose three-fourths of its weight, if raised four thousand miles above the earth's surface. And thus if we take the simplest and most usual test of quantity, instead of being wholly invariable, it varies with every change of elevation above the earth's suface, or even of latitude alone.

Localized force, attractive and repulsive, is more enduring than the secondary qualities of matter, and has the best claim to be regarded as constituting its essence. This is the first lesson of real science. Are these forces, then, fixed and constant, or do they vary? It is a second lesson of science, that they vary continually. They increase as the distances lessen, and they grow less as the distances increase. A pound of gold or lead lessens in weight, though imperceptibly to our senses, every yard that we lift it in the air, every fathom that we sink it in a mine, every mile that we travel with it southward on the earth's surface. If quantum of weight be the essence of matter, a part of that essence is destroyed or created anew by every change of place it undergoes. So far is it from being constant and invariable, it is a function of distances that are ever changing.

But at least we are assured that its reduction to nothing is inconceivable. It is an *à priori* law of human thought. And how is this proved? By thinking only of compression. To try whether force, the subtle essence of matter, can be annihilated, we are instructed to think of those changes only, in which all experience, popular or scientific, conspires to shew that forces are increased.

The true test is plainly just the opposite. We must travel north and not south, to determine whether the North Pole can possibly be reached. Let us try the effects of expansion. Turn a pound of ice into water, the water into vapour, and its solid, sensible resistance to the touch has all disappeared. We test its continued existence less by resistance than by weight. Remove it next four thousand miles upward, and three-fourths of its weight is gone. Transfer this vapour, again, a hundred and forty thousand miles further towards the sun, and its terrestrial and its solar weight will each be five grains only, and, being in opposite directions, will neutralize each other. Remove it beyond the orbit of Neptune, and its solar weight, its most enduring attribute, will be only one two-hundredth part of a single grain. The very test, to which the appeal is made to prove matter indestructible, proves it the reverse, if once we recognize the action of an infinite power. Its essence, to the scientific mind, consists of localized forces, so that force and extension are its two components. Assume an infinite force of compression, and the universe, resisting in vain, would shrink into a mathematical point, and so disappear. Assume indefinite expansion, a process easier to conceive, since we have examples of it ever before our eyes, and the parts of the universe, receding from each other, would melt away into infinite space. It would be like a gigantic bubble, which bursts by expanding, and so finally disappears.

III. But a still more conclusive argument for the same doctrine is endeavoured to be drawn from the very nature of thought itself. It is thus explained:—

"Our inability to conceive matter becoming non-existent is immediately consequent on the very nature of thought. Thought consists in the establishment of relations. There can be no relation established, and therefore no thought formed, when one of the related terms is absent from consciousness. Hence it is impossible to think of something becoming nothing, for the same reason that it is impossible to think of nothing becoming something—the reason, namely, that nothing cannot become an object of consciousness. The annihilation of matter is unthinkable for the same reason that the creation of matter is unthinkable, and its indestructibility thus becomes an *à priori* cognition of the highest order. It is not one that results from a long-continued registry of experiences, gradually organized into an irreversible mode of thought, but one that is given in the form of all experiences whatever."

This reasoning, if really sound, is very wide-reaching and comprehensive in its results. It will prove that there is no such thing in the universe of thought as a beginning or end of anything whatever, substance, attribute, or appearance. The maxim, whatever is is right, is replaced by another—whatever is is eternal, without either beginning or end. What I see now I have always seen. What I feel now I have always felt. And I must continue to see and feel them for ever, from a necessity involved in the very nature of thought. For we think in relations. To think of a visual act as beginning or coming to an end is to compare its present existence with its past or future non-existence. But non-existence is "absence from consciousness." We can establish no relation, frame no thought, of which existence is one term and non-existence the other. Hence it is impossible to think of my present act of vision becoming non-existent, for the same reason that it is impossible to think of its having begun, namely, that the non-existent cannot be an object of consciousness. The past and future eternity, then, of every present sensation is "an *à priori* cognition of the highest order." It is no result of a long-continued registry of experiences gradually organized into an irreversible law. It is a truth, an *à priori* cognition, "given in the form of all experiences whatever."

Such is the argument, on the strength of which Mr Spencer relies, to exclude the whole Christian Church, and all believers in a creation, from the category of thinkers or reasonable beings. It is really a superlative of metaphysical absurdity, thinly disguised from the author of it by the veil of a singular and unusual phraseology. It is pseudo-thinking of the most extreme kind. In fact, between Sir W. Hamilton and Mr Spencer, thought of every kind must be impossible, and eliminated from the universe altogether. According to the first, we cannot think of the unconditioned or unlimited, for this is to transcend consciousness, which deals only with contrast and limitation. According to Mr Spencer, we cannot think of the conditioned and limited, for this is to think of a contrast between existence within the limits, and non-existence before or after or beyond them, and thought with non-existence for one of its terms is impossible. The doctrine includes a patent of past and future immortality, not only for every insect, but for every transient and momentary sensation. The Theist, the Pantheist, and the Atheist have already been convicted of pseudo-thinking for the one truth they hold in common, that "there must be self-existence somewhere." The rival creed there propounded, as the test of philosophical orthodoxy, is that self-existence is inconceivable and unthinkable anywhere. And here we have a second substitute, that there must be self-existence everywhere and in everything, and that any different creed is impossible from the very nature of thought. No true philosophy can ever be founded by such violent plunges into the depths of opposite absurdities.

An explanation is offered, how so large a part of mankind should have fancied themselves to believe the exact opposite of an *à priori* truth "not only of equal certainty with those commonly so classed, but of even higher certainty." This most certain of all *à priori* truths is really the most strange of all conceivable falsehoods, that there has never been, and could never be, either a beginning or end of anything whatever, substance, attribute, phenomenon or sensation, but that each and all, by the very nature of thought, is from everlasting and to everlasting. Mr Spencer writes as follows: "To set down as a proposition which cannot be thought, one which mankind once universally professed to think, and which the great majority profess to think even now, seems absurd. The explanation is that in this, as in countless other cases, men have supposed themselves to think what they did not think. The greater part of our conceptions are symbolic. Many of these, though rarely developed into real ones, admit of being so developed, and being directly or indirectly proved to correspond with the realities, are valid. But along with these others pass current, which cannot be developed, or by any direct or indirect process be realized in thought, much less proved to correspond with actualities. The legitimate and illegitimate are confounded together; and, supposing themselves to have really thought what they have thought only symbolically, men say they believe propositions of which the terms cannot even be put together in consciousness. Hence the ready acceptance given to sundry hypotheses concerning the origin of the universe, which yet are absolutely unthinkable."

Error may be of two different kinds. We may mistake the possible for the actual, supposing something to have really occurred, which is disproved on exact inquiry. Or we may confound with essential truth what is essentially false, and involves a secret self-contradiction. And this second kind is plainly more aggravated and dangerous than the first. No rule or law can be laid down, by which we can escape from the risk of so great an evil. Whoever could invent a specific by which to discern infallibly truth from falsehood, an illegitimate from a legitimate conception, might rank among the foremost benefactors of the human race. The lesson which comes nearest to this character, has been given us by Him who is the Supreme Wisdom and Perfect Truth :—"If thine eye be single, thy whole body shall be full of light." But the doctrine which condemns the whole Church of Christ, and the great body of mankind, as guilty of self-deception and folly, because they have held either the creation or the annihilation of matter to be possible in itself, and the first of these to be a fact divinely revealed, and rests its condemnation on a maxim contradicted by the experience of every hour, in each individual of mankind, seems at the furthest possible remove from this high eminence of superior insight and wisdom. The conception denounced as illegitimate, and by the rejection of which creation and annihilation are both disproved, is the notion that anything whatever can either have a beginning of existence or cease to be. This is pronounced impossible by virtue of the universal laws of human thought. The exact reverse is so manifestly true, that it is lost labour to make the truth plainer by a process of reasoning. It is the basis of all experience, the pre-condition of every process of human thought, and has embodied itself in the vocabulary and the structure of all the countless languages of mankind.

CHAPTER VII.

§ I. THE CONTINUITY OF MOTION.

THE Indestructibility of Matter, the Continuity of Motion, and the Persistence of Force, are three main premises, on which the Doctrine of Physical Fatalism is founded. But since motion is treated as one mode of Force, the second and third of these are hardly distinguishable from each other. The terms, indestructibility, continuity, and persistence, are practically equivalent. One and the same thought pervades the three chapters, that there is a fixed quantum of matter, motion, and force, which may vary indefinitely its form or its locality, but remains unaltered and unchangeable amidst all those changes, which make up the history of the universe.

The first exposition of the Continuity of Motion is in these words:

"The axiomatic character of the truth that Motion is continuous is recognized only after the discipline of exact science has given precision to the conceptions. Aboriginal men, the uneducated, and most of the so-called educated, think in an extremely indefinite manner. Accepting without criticism the dicta of unaided perception, to the effect that surrounding bodies, when put in motion, soon return to rest, the great majority tacitly assume that the motion is actually lost."

The experiments are then mentioned, which have removed the obstacles to the reception of Newton's first law of motion; namely, that a moving body, not influenced by external forces, will remain at rest, or move on uniformly in a straight line. This is an adoption and restatement of Newton's own Scholium, which precedes the First Book of the Principia. It implies that this first law is simply an inductive result of observation and experiment. So Sir W. Thomson and Prof. Tait remark, in their able treatise on Natural Philosophy. These laws, they observe, "must be considered as resting on convictions drawn from observation and experiment, *not* on intuitive perception."

Mr Spencer, however, has no sooner accepted this opinion, and repeated some of the observations on which it rests, than he proceeds to maintain another, the exact reverse; namely, that this law of motion is an *à priori* truth, of which the opposite is unthinkable. He writes as follows:

"The Indestructibility of Motion is not only to be inductively inferred, but is a necessity of thought, its destructibility having never been truly conceived at all, but having always been, as it is now, a mere verbal proposition that cannot be realized in consciousness, a pseud-idea. Whether that Absolute Reality which produces in us the consciousness we call Motion be or be not an eternal mode of the Unknowable, it is impossible for us to say. But that the relative reality we call Motion never can come into existence, or cease to exist, is a truth involved in the very nature of our consciousness. To think of Motion as either being created or annihilated, to think of nothing becoming something, or something becoming nothing, is to establish a relation in consciousness between two terms, of which one is absent from consciousness, which is impossible. The very nature of intelligence negatives the supposition that motion can be conceived, much less known, either to commence or to cease."

We cannot conceive anything that now moves to stop moving, or anything now at rest to begin to move! This is gravely affirmed to be an *à priori* truth, involved in the nature of thought, of which the opposite is unthinkable. When such statements are made in the name of advanced philosophy, to correct the crude pseudo-thinking of half-educated or aboriginal men, we may well lift up our hands in silent amazement. The writer cannot possibly mean what he says. That is an absurdity too great to be held by any reasonable being, much less to be turned into an *à priori* truth. Let us try to ascertain in what esoteric or non-natural sense the words are used. And here the closing sentences of the chapter come to our aid.

"That which defies suppression in thought is really the force which the motion indicates. The unceasing change of position, by itself, may be mentally abolished without difficulty. We can readily imagine retardation and stoppage to result from the action of external bodies. But to imagine this is not possible, without an abstraction of the force implied in the motion. We are obliged to conceive this force as impressed, in the shape of reaction, on the bodies that cause the arrest. And the motion that is communicated to them we are compelled to regard, not as directly communicated, but as a product of the communicated force. We can mentally diminish the velocity, or space-element of motion, by diffusing the momentum, or force-element, over a larger mass of matter. But the quantity of this force-element, which we regard as the cause of the motion, is unchangeable in thought."

The unceasing change of position, to borrow Mr Spencer's phrase, in this new philosophy of Physical Evolution, makes its examination very difficult. The Proteus has no sooner been grasped in one shape, than he slips from our hands, and stands before us in another, wholly different. In the present instance, three paradoxes, contradicting each other, and more and more contradictory to reason and common sense, are successively laid down as grand *à priori* truths. The climax is reached in the assertion that it is impossible to conceive of anything either beginning or ceasing to move. But then follows a speedy admission that such a conception, which indeed is one of the most frequent and habitual experiences of human thought, is quite easy and natural. And still this retractation of an outrageous paradox is accompanied with two or three fresh errors of a very radical kind. Let us trace the steps of the reasoning in succession.

First of all, Theism, Pantheism, and Atheism are condemned and rejected for one common fault. They agree to assume " self-existence somewhere." But this assumption, whether made nakedly or under a disguise, is vicious and unthinkable. It is one which we cannot avoid making, but still we ought to avoid it. And the special excellence of Religious Nihilism is that it does what its advocate says cannot be done, and refuses to admit selfexistence anywhere, because self-existence is unthinkable.

Such is the first paradox of the system, and a greater outrage on sound reason is difficult to conceive. But we pass on to the "Indestructibility of Matter," and meet there with a second paradox, the exact contrary and antithesis of the first. The annihilation of Matter, we are told "is unthinkable for the same reason that the creation of matter is inconceivable." It contradicts the very nature of thought. "It is impossible to think of something becoming nothing, or nothing becoming something, for the same reason, namely, that nothing cannot become an object of consciousness."

Here, then, it is pronounced to be a contradiction of the laws of thought, that anything should either begin or cease to be. Theism is first coupled with Pantheism and Atheism, and condemned to death and burial as a deceiver of mankind, because it affirms self-existence somewhere, while self-existence is inconceivable. And next we are taught the exact converse, that self-existence is the only kind of existence conceivable. Whatever exists now must always have existed, and must exist for ever; since it is forbidden by the very nature of thought to think of anything whatever as either beginning or ceasing to be.

A third and still greater paradox follows. What has been affirmed of Matter, that it is self-existent by the very nature of thought, is next affirmed of Motion also. This too can never begin or cease. For this would be to establish a relation in consciousness between two things, one of which is not in consciousness, that is, motion, and not-motion or rest. Motion is being, rest is not-being, not-being is nothing, nothing is not in consciousness, and cannot even be thought of. Change of place is thus treated as no mere attribute of bodies, or relation between them, but a real substantial thing. Hence it shares the prerogatives just assigned to Matter. It is uncreated, self-existent, and eternal.

Motion, in this theory, seems to be conceived as a liquid, and moving bodies are like solid, empty vessels, capable of receiving it. The liquid may be poured from one vessel to another. It changes its own place, motion itself is moving, every moment. It is transferred from vessel to vessel with the speed of lightning, from atom to atom, and from star to star. But its total quantity remains invariable. Such as it was in the beginning, such is it now in total amount, and must continue the same in amount for ever.

This is indeed a symbolic conception of the illegitimate kind. It is equally opposed to the notions of uneducated men, and the ripest conclusions of Dynamical Science. For the collective motion of a system can only be constant, when the Potential Energy, or integral of the force, is constant; and whenever the forces are functions of the distance, as in the case of universal gravitation, and of molecular cohesion and repulsion, this must always vary with variations of the mutual distances. The effects of condensation may be partially compensated by expansion in other parts of the system. But a total compensation, under the law of gravitation, or any similar law, is dynamically almost impossible.

But after we have reached this extreme of error, which makes motion a real substance, not an attribute or relation of substances, and self-existent, without beginning or end, we return once more to the region of common sense. Motion *per se*, "the unceasing change of position, considered in itself," it is finally admitted, may be mentally abolished without difficulty. The metaphysical proof that the conception itself is impossible, because it implies a relation in thought between something and nothing, is discarded as soon as it has been advanced, and a wholly different statement replaces it. We can easily conceive a body either to begin or cease to move. But we cannot conceive this to occur without there being some cause of the change.

This is perfectly true. The usual definition of Force is that which produces or tends to produce motion, or a change in the amount or direction of motion. Sir Isaac Newton, Sir W. Thomson, and Professor Tait, agree in holding the first law of motion to be an inductive conclusion from observation and experiment. Mr Spencer, we have seen, professes to share this view, but to hold along with it one wholly opposite, that it is a necessity of thought, an *à priori* truth. I believe the correct view to be that it is simply a definition of force, that is, of Physical Force, the subject of Dynamical Science. What experiment can do is not to prove the action of some force, when velocity is increased or diminished or changes its direction. That is proved by the change itself. Experiment serves only to define and localize it, to connect it with the presence, and the nearness or distance, of other bodies; as the air, through which a bullet passes in its course, or the sun, as the deciding cause and main seat of that force which determines the planetary motions.

The final shape of this doctrine of the Continuity of Motion consists in an exposition of some necessary process of thought, when we think of motion as transferred. The elements of necessary thought, we are told, are these. First, force is implied in the motion of the striking body. Next, this force is impressed in the shape of reaction on the body struck, or that causes the arrest. Thirdly, we cannot conceive the motion of the struck body as directly communicated, but as a product of the force which has been thus communicated. Fourthly, velocity is the space-element of motion, and momentum the force-element. Lastly, we can diminish the first, or space-element, by diffusing the force-element over a larger mass, but this last element we cannot conceive to have its quantity varied. It is unchangeable in thought. Of these five statements, or *à priori* principles of the philosophy in question, the first is ambiguous and misleading, and all the others are untrue.

First, force is implied in the motion. This is true in the sense that it is implied as a previous cause. But it is not implied as a present existence, since the essence of the law is this, that uniform motion in a right line proves not the present action of a force, but its absence. It is implied further that if the motion is to cease, there must be a future action of force, to cause that cessation. But present force in the motion itself is not implied, but negatived, by Newton's first law, which forms the very starting-point of all the principles of Dynamical Science.

Next, this force is impressed in the shape of reaction on the body struck, and thus communicated to it, and the motion of the struck body is the product of the force thus communicated.

This is completely untrue. The force exercised by A in striking B is attended with a reactive force of B on A. But this second force does not cause the motion of B, but the first force only. And that first force is not dependent directly on the motion or speed, but simply on the distance. It is insensible at sensible distances, but rapidly increases at the distance of seeming contact. The reaction of the second body is wholly employed in arresting the motion of the first body, and not in causing its own. The notion that a force is first transferred from A to B, and then, residing in B, causes B to move, contradicts the fundamental conceptions of Dynamics. It is a case of pseudo-thinking and mere illusion.

Again, velocity is not the space-element of motion,

and momentum the force-element. Momentum is simply the product of the velocity by the mass, that is, in other words, the sum total of the velocities of all the parts or atoms of the moving body. When a product is constant, of course the increase of one factor requires the lessening of the other. But a sum of velocities can have no nearer relation to force, and no less intimate relation to space, than the velocities of which it is the sum.

That the momentum, in collisions, is unchangeable in thought, is a further error. It is not even unchangeable in fact, except under special dynamical conditions. In fact, Sir Isaac Newton, in his Optics, looks on the opposite view as certain, and writes on it as follows:—

"From the varying composition of two motions, it is very certain that there is not always the same quantity of motion in the world. For if two globes, joined by a slender rod, revolve about their centre of gravity with a uniform motion, while that centre moves uniformly in a straight line in the plane of their motion, the sum of their motions, when they are in the right line described by their centre of gravity, will be greater than the sum when they are in a line perpendicular to that line. By reason of the tenacity of fluids, and want of elasticity in solids, motion is much more apt to be lost than got, and is always on the decay. If two equal bodies meet directly in vacuo, they will by the laws of motion stop where they meet, and lose all their motion, and remain at rest, unless they be elastic....Seeing therefore the variety of motion we find in the world is always decreasing, there is a necessity of conserving and recruiting it by active principles, such as are the cause of gravity, and the cause of fermentation. For we meet with very little motion in the world besides what is due to these active principles."

The invariable constancy of momentum, or quantity of motion, is thus far from being any necessity of human thought. The greatest of all physical discoverers, at a time when his latest discoveries were made, held a view exactly opposite, that motion is in a state of continual decay. The grounds upon which he rested this opinion have been modified, and to some extent set aside, by later discoveries. The notion of direct collision between finite atoms has received no confirmation, and the tendency of research has been to develope the range of the "active principles" mentioned by Newton in this passage, and indirectly to confirm that hypothesis of force-centres, which Mr Spencer rejects as unthinkable, but on which the conservation of *vis viva*, the only scientific form of the persistence of force or continuity of motion, seems really to depend. And while here the principle of the continuity of motion is laid down as an *à priori* truth, in a later part of the work the exact opposite is maintained, that evolution must issue in equilibration, and equilibration end in complete rest. That such a state must be the outcome of the processes everywhere going on seems, it is said, to be beyond a doubt, and the natural inquiry accompanies the remark—Are we not then manifestly progressing towards omnipresent death? So inconsistent with itself is the new philosophy, in every stage of its progressive evolution!

§ II. THE CONSERVATION OF FORCE.

The Conservation of Force is another main pillar of the modern theory of Physical Fatalism or Evolution. It has been said to bind nature fast in the bonds of fate to an extent not before recognized, to bring vital as well as physical phenomena under its dominion, and to be an idea of the widest grasp and most radical significance. It seems, in fact, to hold nearly the same place in the doctrine of Evolution, which the Being and Perfections of God occupy in Christian Theology. I shall here confine myself to the examination of its meaning and nature, as expounded in the sixth chapter of Mr Spencer's Doctrine of the Knowable. I have elsewhere treated of the subject on its dynamical side.

And first, Mr Spencer would replace the word Conservation by Persistence, in order to avoid the suggestion of a Preserver, and an act of conserving. But the change is needless, and even injurious. Even if Physical Force were the sole divinity, and any implied reference to an All-wise Creator and Preserver should be shut out, as inconsistent with our improved philosophy, it is surely quite easy to expound the conservation of force as meaning self-preservation. On the other hand, Persistence is not free from a moral implication, and one of an unpleasant kind. It means, as dictionaries tell us, "perseverance in a good or evil course, usually in one injurious, obstinacy or contumacy." So a loquacious person persists in talking, when he ought to keep silence. And thus the new term may free us from the risk of admitting the presence of a Divine Preserver. But it suggests instead the idea of a deaf, blind Fate, which persists in acting without any reasonable motive, heedless of all obstacles, and brooking no control from either human or Divine intelligence. If a term were desired, free from any moral implication, perhaps the Inherent Permanence of Force would be the best.

What, now, is the place assigned to this doctrine in the wide circuit of human science? According to Mr

Spencer it is the first and highest of all truths, the most certain, and also the most important. He speaks of it in these words:

"The Persistence of Force is an ultimate truth, of which no inductive proof is possible.... Whoever contemplates the relation in which it stands to the truths of science in general will see that the truth transcending demonstration is the Persistence of Force. It is not only a datum of science, but even a datum which the assertion of our nescience involves. Deeper than demonstration, deeper than even definite cognition, deep as the very nature of Mind, is the postulate at which we have arrived. Its authority transcends all other whatever; for not only is it given in the constitution of our own consciousness, but it is impossible to conceive a consciousness so constituted as not to give it. The sole truth, which transcends experience by underlying it, is thus the Persistence of Force. This, being the basis of experience, must be the basis of any scientific organization of experience. To this ultimate analysis brings us down, and on this a rational synthesis must build up." (F. P. 188—191.)

Our first inquiry must be—What is the Force intended in this great doctrine? and our second, What is meant by its persistence?

The nature of Force is thus explained. It is not the force we are immediately conscious of in our own efforts, for this does not persist. Hence the Force, of which we assert persistence, is that Absolute Force, of which we are indefinitely conscious as the necessary correlate of the Force we know. By the persistence of force we mean really "the persistence of some Power which transcends our knowledge and perception." "In other words, the assertion is another mode of asserting an unconditioned Reality without beginning or end." "The truth that Force is indestructible is the obverse of the truth, that the unknown cause of the changes going on in consciousness is indestructible." "The Persistence of the universe is the persistence of that unknown Cause, Power, or Force, which is manifested to us through all phenomena."

Next, what is meant by its persistence? To this question a distinct answer is repeatedly given. It is the constancy or sameness of its total amount, amidst ceaseless changes in its locality or distribution. We read as follows:

"The perceptions (from which science infers that Matter is indestructible) amount simply to this, that the *force* which a given quantity of matter exercises remains always the same. This is the proof on which common sense and exact science alike rely. The obvious postulate is, that the quantity of matter is finally determinable by the quantity of gravitating force it manifests." "Where, as in celestial physics, the continuity of motion is quantitatively proved, the proof is not direct, but inferential, and forces furnish the data for the inference.... The quantity of the force-element, which we regard as the cause of the motion, is unchangeable in thought.... That the quantity of Force remains always the same is the fundamental cognition, in the absence of which the derivative cognitions, *i. e.* the indestructibility of matter, and the continuity of motion, must disappear." "Every antecedent mode of the Unknowable must have an invariable connection, quantitative and qualitative, with that mode of the Unknowable which we call its consequent." "It is impossible to conceive the product of joint action in one case as unlike that in the other, without conceiving one or more of the forces to have increased or diminished in quantity; and this is conceiving that Force is not persistent."

This great *à priori* truth, greater than all other truths, and transcending both demonstration and experience, proves thus at the outset to be nothing else than an open and direct self-contradiction. The Force intended in the phrase is not the phenomenal, but the real. It is the Unknowable and Immeasurable. It is the same, of which we have been previously assured, as the "widest, deepest, and most certain of all truths, that it is utterly inscrutable." And now Persistence of Force is a second truth, which shares the same prerogatives. It transcends demonstration, and both transcends and underlies experience. And its persistence means constancy of its measured value, strict equality, without variation, in its total amount. We can know, with the highest certainty conceivable, that something, of which it is the deepest of all truths that we can know nothing at all, is measurable. We can measure its separate parts, and compare them. We can sum up these measurable parts into a total, also measurable. We can compare the values of this total from moment to moment or from age to age. Finally, we can complete the fundamental axiom, that nothing whatever can possibly be known of this Absolute Force, by a second axiom, equally fundamental, that its total amount is finite and measurable, and that this total never varies, and is incapable of the least diminution or increase to all eternity.

Such is the doctrine of the Persistence of Force in its own nature. If we accept the double definition of Force and Persistence in the First Principles, it is a direct self-contradiction. On what grounds is it affirmed to be an *à priori* truth?

The process of reasoning is remarkable. First, it cannot be proved *à posteriori*. "We cannot infer the persistence of force from our own sensation of it, which does not persist. We must infer it from the continuity of Motion, and the undiminished ability of Matter to produce certain effects. But to reason thus is manifestly to reason in a circle. It is absurd to allege the indestructibility of Matter, because under whatever change of form a given mass exhibits the same gravitation, and then to argue that gravitation is constant, because a given mass of matter exhibits always the same quantity of force. Quantitative science implies measurement, and measurement implies a unit of measure.......Everything turns on the truth of the assumption, that the gravitation of the weights is persistent, and of this no proof is assigned, or can be assigned......No problem of celestial physics can be solved without the assumption of some unit of force. The validity of one or other inference depends wholly on the assumption that the unit of force is unchanged......Clearly, then, the persistence of force is an ultimate truth, of which no inductive proof is possible."

This is the main proof that the Persistence of Force is the first of all *à priori* truths, highest in dignity and importance. First, its truth is postulated or assumed. Next, it is shewn that an inductive or *à posteriori* proof is impossible, because no experience can prove the constancy of the unit assumed. Since, then, it is certainly true, and cannot be proved inductively, it must be an *à priori* truth. By such a process it is plain that any falsehood whatever may be promoted to the same high dignity of an *à priori* truth.

A second proof, however, is offered in connection with a new and altered definition of the doctrine. "The equality of action and reaction is taken for granted from beginning to end of either argument;" that is, in the reasonings both of terrestrial and celestial physics; and "to assert that action and reaction are equal and opposite is to assert that force is persistent."

Now the equality of action and reaction is Newton's third law of motion. He there occupies four pages with experimental confirmations of its truth. Our latest authorities in Dynamics, Sir W. Thomson and Prof. Tait, in their able treatise on Natural Philosophy, take the same view, that this equality is an *à posteriori* result of observation. That an opposite view is not inconceivable admits of a very simple proof; for Prof. Bayma, in his Molecular Physics, adopts an hypothesis inconsistent with it, and develops its consequences in three hundred pages of dynamical reasoning.

Again, the persistence of force, in its dynamical sense, is not the same with the equality of action and reaction, but wholly distinct. Take the case of two bodies, mutually attractive, falling towards each other from a state of rest. Here action and reaction are equal and opposite. The first pulls the second as much as the second pulls the first, and in an opposite direction. Yet neither the forces nor the motions are constant, but go on increasing together, till their amount is infinite at coalescence. The second argument, then, rests on two premises, which are both untrue; that Newton's third law is the same with the persistence of force, and that it is an *à priori* truth.

The doctrine, indeed, in Mr Spencer's work has five different meanings. First, it is Newton's third law. "To assert that action and reaction are equal and opposite is to assert that force is persistent" (p. 188). This is a truth, but one of induction, and quite distinct from the conservation of energy, the dynamical equivalent of this persistence of force. Secondly, it is the same with the non-annihilation of matter, and means that "the force a given quantity of matter exercises remains always the same" (p. 177). This double identity is doubly untrue. The force a given quantity of matter exercises is so far from remaining the same, that by the law of gravitation, and every other known or conjectured law, it varies with every variation of mutual distance. The constant of force, so far as we know, does not vary. But, under the atomic theory, or the doctrine of force-centres, it might vary, and no single atom be annihilated. An atom would not be annihilated, if it were to attract and repel only half as much as it does now. There are thus four errors in this one statement. The persistence of force is not the same with the non-annihilation of matter. The non-annihilation of matter is not the same with the invariability of the force exercised by each separate atom, and invariability is not a fact or truth at all, but exactly the reverse, that the exercised force varies every moment.

Thirdly, the doctrine is said to mean the constancy of each force in any system of forces; for "to conceive one or more of the forces to have increased or diminished is conceiving that force is not persistent" (p. 193). In this sense it will be directly opposed to all experience, and every case of dynamical reasoning. For all these imply and require the variation of individual forces.

Fourthly, it is the constant variation of all forces, attractive or repulsive, by the law of the inverse square of the distance, and no other (p. 60). For this, we are told, is no empirical law, but deducible mathematically from the laws of space, and its negation is inconceivable. This fourth meaning is the reverse and contradiction of the last. It is also a direct contradiction of the Principia and of all dynamical science, and all experience. For in this case there could not possibly be both attraction and repulsion, but one of these alternatives alone.

Lastly, it is "the persistence of some Power that transcends our knowledge and conception, an unconditioned Reality, without beginning or end" (p. 189). In this sense, it will be a philosophical substitute for the Christian doctrine of the self-existence of God, with this all-important contrast, that the Supreme and Absolute Power is recognised as Power alone, but denied all cognisable moral attributes or perfections whatever. It is the consecration of all tyranny, and the apotheosis of destiny, placing blind, irresistible Force, without reason, choice, wisdom, or moral goodness, on the throne of the universe. And persistence must also change its meaning. It can no longer be constancy of amount. For quantity is a condition, and the Unknowable is the Unconditioned. Equality in amount implies measurement, and there can be no measurement of the Unmeasurable.

The Persistence of Force, then, as taught in the First Principles, is wholly ambiguous. In one of its five meanings it is a familiar truth of Dynamics, but a truth resulting from observation and experiment. In each of the four other meanings it is untrue and self-contradictory. But the stress which has been laid upon it, and the confidence with which it is propounded as a grand discovery, with the vastness of superstructure which it has been endeavoured to rear upon it, are reasons for carrying the analysis a little further, and examining the other statements in the short chapter of Mr Spencer's work, which deals with the subject.

The first of these is an assertion that the truth of this doctrine is indispensable to the existence of Science.

"An attempt to ascertain the laws to which manifestations in general and in detail conform would be absurd, if the agency to which they are due could either come into existence or cease to exist. The succession of phenomena would in such case be altogether arbitrary, and Science, equally with Philosophy, would be impossible."

Science and Philosophy would be impossible, and the succession of phenomena wholly arbitrary, if there were no settled laws of material action and physical change. If all changes, or a large number, were like the supposed deflections of the atoms of Epicurus, capricious and causeless, there could be no room for science. But a past creation of the world out of nothing, and even its future return, at some distant age, to nothing again, would not interfere at all with the reality of science. They would simply confine its range within the limits of actual time, past or future. The second hypothesis, however, implying the reversal and abrogation by the Creator of his own work, is open to moral objections of the most grave and vital kind. But these do not apply to the doctrine of Creation. It simply replaces the notion of a blind, unaccountable, irresistible Fate, by the conception of an orderly scheme of Providence, secretly dependent throughout on the will and purpose, the choice and counsel, of a Being supremely great, and perfectly good and wise.

In like manner, suspensions or modifications of the results of physical laws by the will of the Lawgiver who appointed them, to fulfil some end of his moral government, could not destroy science, or reduce its conclusions to entire uncertainty. They would merely make their truth and realization depend on one further condition of a higher kind. Law, in its highest and truest sense, would remain supreme. But physical laws would be seen to be instruments of the Divine government, which must surely have some nobler object than to vary the co-ordinates in space of a vast number of pulling and pushing atoms, destitute of any higher gift or power than change of place alone. They will be seen in their true light, as handmaidens of that higher Law, "whose seat is the bosom of God, and her voice the harmony of the world." It is chance and mere caprice, not Moral Law and Divine Sovereignty, which alone could interfere with the reality of Physical Science.

The doctrine is next defined to be one of quantitative measurement, while yet we can have no experimental proof that all the forces do not vary.

"In all three cases the question is one of quantity. Does the matter or motion or force ever diminish in quantity? The quantity of matter is asserted to be the same, if the number of units of force it counterbalances be the same. The validity of the inference depends entirely on the constancy of the units of force. In the reasonings of the astronomer there is the like implication. No problem in celestial physics can be solved, without the assumption of some unit of force."

That such an argument entirely fails to fulfil its own object, and prove that the invariability of force is an *à priori* truth, must be self-evident. But the subject will repay a little further investigation.

What is true in the argument is the plain fact that force and motion have no natural unit. To measure and compare them, a unit needs to be assumed. The same is true of space in all three dimensions. There is no natural unit of length. It is true also of time. But if we assume a unit of length and of time, we may derive from these a natural unit, both of velocity and of accelerating force. And thus, subject to the double assumption, both forces and velocities will be truly represented by geometrical lines of varying length.

Now what is the true inference from such a premise? Is it the persistence of force or motion, or the constancy of some total, formed by summing up all the separate forces or motions of the universe? It is plainly the exact reverse. A total formed of such elements cannot possibly be either self-existent or invariable. Each of the elements that compose it is varying every moment. It is truly represented by a line ever varying in its length. The grand total, every moment, will be the sum of $n.n-1$ elements, and truly represented by the sum of $n.n-1$ geometrical lines, where n is the number of atoms in the universe. The Persistence of Force and the Continuity of Motion, as expounded in First Principles, will thus resolve themselves into a paradox of this amazing kind. The Power which the universe manifests is utterly inscrutable. To suppose that we can know anything concerning it, or fitly ascribe to it personality, will, goodness, wisdom, is one of the countless impieties of the pious. But this we may know concerning it, that it is truly represented by a finite straight line of definite length, which is made up of as many parts as there are pairs of atoms in the universe, and of which every part varies perpetually by laws mainly unknown to us, while the finite length of the total remains the same for evermore. The golden calf was a respectable idol, compared with this philosophical substitute for the True and Supreme Reality.

One further remark remains to be examined. "The Force," it is said, "of which we assert persistence, is that Absolute Force of which we are indefinitely conscious, as the necessary correlate of the force we know."

In this sentence the confusion of thought, involved in the whole theory, comes out into very clear relief. This doctrine of Persistence asserts a quantity of force, unchangeable amidst ceaseless changes of form and distribution. Now quantity is a condition. What the doctrine, then, affirms, is the sameness of one main condition, quantity, in the Unconditioned, a measurable sameness in the Immeasurable.

Again, the Force we know, and of which we are conscious, and not a force of which we can know nothing, is the only Force of which we can consistently affirm anything. But the force we are conscious of, Mr Spencer expressly affirms, does not persist (p. 195, l. 5). By his admissions, then, the non-persistence of knowable force, the only force knowable, is a certain truth; while persistence and non-persistence are predications equally valid, equally incapable of being proved or disproved, when we speak of the Absolute Force, which is unknowable.

The source of all this error and confusion arises from mixing together two steps in the healthy and normal development of thought, and confounding them into one. The simple contrast laid down is of the Phenomenal and the Real, the Knowable and the Unknowable. But the real contrast is twofold, of phenomena or sensations, which are momentary and evanescent, and of material objects, and individual minds or persons, which abide and endure; and again of things and persons, local, limited, and finite, with the Supreme and Absolute Reality on whom they all depend.

A sensational philosophy seeks to get rid of things and persons, speaks of them as wholly unknowable, and strives to create and build up a universe and a science of sensation and appearances alone. Religious Nihilism would leave the Great, Supreme Reality, the I AM of revealed religion, in the sphere of existence, but shut Him out for ever, for all mankind, from the sphere of knowledge. But all the genuine discoveries of physical and mental science utter one ceaseless protest against this double error. They have been made by passing from the phenomenal to the real, and yet to the finite, not the Supreme Reality. They refer, not to sensations, but to the causes of sensations, to localized and measurable forces, and to the powers, gifts, and faculties of individual minds. Yet these, as localized and measurable, or self-conscious, finite, and individual, stand in plain contrast to the First Cause, the Absolute and Infinite Being.

To obtain permanence at all, we must bid adieu to fleeting phenomena which do not persist, and look higher. And thus, when no other contrast is admitted but of the Knowable and the Unknowable, of mere appearances and an Absolute Reality behind appearances, the second of these, and not the first, must be the home of genuine science. But the doctrine of indestructibility, continuity, and persistence, is one of measurement, of equal values and amounts of matter, motion, and force. Now this is wholly irreconcilable with the doctrine of the Unknowable. And thus there is an uneasy and perplexing oscillation, where no footing, on the principles of the theory, can be found for genuine science on either side. It seeks a home among phenomena; but phenomenal matter is destroyed continually. Phenomenal motion can and does cease, and can be mentally abolished without difficulty; and the force of which we are conscious in our own efforts does not persist. We turn, then, to the reality, of which the phenomena are manifestations. And here we are met with the equal difficulty, that by the doctrine of the new philosophy, this is wholly inscrutable. Science can find no solid resting-place, till it rises beyond the ever-shifting phenomena, and occupies itself with Real Being, in the three successive gradations, real matter, real mind, and a Real Author of mind and matter, the Absolute Being, or great First Cause, the Father of Spirits, the Creator, Preserver, and Governor of the moral and material universe.

CHAPTER VIII.

THE TRANSFORMATION OF FORCE AND MOTION.

The Persistence of Force, as held by Mr Spencer, who makes it a sequel of the doctrine of the Unknowable, and a great *à priori* truth, involves a direct self-contradiction. Phenomenal Force does not persist. Force not phenomenal, or the reality which lies hid behind appearances, by the dicta of this philosophy, is wholly beyond the range of human knowledge. To affirm either that it does or does not persist involves the same reversal and abandonment of the negative creed, as those theological dogmas which its advocate so strongly condemns.

The same doctrine, however, under the name of the Indestructibility of Force, or Conservation of Energy, may be held, apart from the doctrine of the Unknowable, as an inductive result of modern physical research. Such seems to be the view taken by Mr Grove in his Correlation of Forces, and Dr Tyndall in his Belfast Address; while Mr Spencer seeks, as best he may, to combine it with his own higher claim, that it is an *à priori* truth, and incapable of being established by induction and experiment. Many, who reject the opinion that it is a necessary truth, like those of pure arithmetic and geometry, look upon it as a grand result of recent discoveries. It is necessary, then, to examine closely what it means.

Three main results of modern physics have modified opinions which had prevailed before, and constitute the nucleus of scientific truth in this doctrine of the Conservation of Energy. The first is the establishment of the Baconian view of heat, as intestine or atomic motion; in contrast to the later doctrine, embodied in most works of the last century, and in the treatises of Fourier and Poisson, where it is assigned to a distinct fluid of caloric, supposed to permeate all the parts of solid, liquid or gaseous bodies. An easy inference from that theory is the equivalence of a certain amount of heat with a certain amount of sensible motion; and this has been fixed and ascertained by the experiments of Seguin, Joule, Sir W. Thomson, and others.

A second result of recent research removes the partitions, by which the theories of Light, Heat, Electricity, Magnetism, were severed wholly from each other, and referred to five or six different fluids or imponderables. The chief alternatives now recognised are that they result from the interaction of matter and one or two kinds of ether, or from the action of matter alone. The two extreme views have their scientific advocates. But the more common, and I believe the juster view, is that one kind of ether, besides gravitating matter, is required, and one only, to explain the whole range of the phenomena of material change.

But the third element is the most important, and constitutes the essence of the doctrine as a principle of Dynamics. It may be thus briefly explained.

Assume a finite system, however immense, of atoms, either extended or unextended, which are endued with attractive or repulsive forces. Assume further that these atoms never touch, so as to stop each other's motion by their mere solidity or impenetrability. Let their motion increase or diminish, only from the action of attractive or repulsive forces, and these forces be all functions of the mutual distance. Another function, the integral of the first, is called the Potential Function. The difference between two values of this function for two different sets of places of the atoms will express the total amount of force exercised on those atoms, in the passage from one position to the other. The effect of these forces, so exercised, will be to increase or diminish the velocities. Now if we take the half square of the velocity for the measure of the motion of each mass or atom, then the amount added to or taken from the collective motion will exactly equal the positive or negative amount of the forces exerted by the atoms or masses on each other during this interval of time.

The value of the integrals of the laws of force, for any set of positions, is called the Potential Energy of the system in that position. The difference of two such values is the gain or loss of Potential Energy, in passing from the first position to the second. In like manner the sum total of half the squared velocities of all the atoms or masses at any moment of time, is called the *Vis Viva* or Kinetic Energy of the system at the moment. The difference of two such values for two successive positions will be the gain or loss of *vis viva* or kinetic energy.

The doctrine of the Conservation of Energy is the assertion that the sum of the two kinds of energy, potential and kinetic, is always constant, so that as much is gained by the one as is lost by the other. Hence energy comes to be looked upon as an indestructible something, which may change its form or its seat continually, but of which the total amount is fixed and invariable. In the doctrine of evolution it becomes a physical Demiurge, and is supposed able to account for all known or unknown changes, without leaving room for any action of mind, or the intervention of a Creator and Moral Governor of the world.

To see the true meaning, however, more clearly, three cases must be separately considered. The first is that of purely repulsive forces; the second, of forces purely attractive; and the third, of forces which change from repulsion at the smallest distances to attraction at all others, with a neutral limit between them.

A Repulsive Potential, when the repulsion varies by any inverse powers of the distance beyond the first, is plainly finite in value for a finite distance, and zero when the distance becomes infinite. The effect of the repulsion also is to separate the masses or atoms from each other. Thus, as the velocities increase, the Potential decreases. To make their sum constant, the Repulsive Potential must have a positive sign.

The value of an Attractive Potential is also finite at a finite distance, zero at infinity, and infinite when the distance vanishes, if the attraction varies as the inverse square, or by any inverse power beyond the first. But the amount thus increases as the bodies approach, when the velocity also will increase. That the sum of the Potential Function and the *vis viva* or Kinetic Energy may be constant, the first must be reckoned negative. Thus the algebraic sum is the real difference. Supposing the system to have started from a state of rest, what is really constant is the excess of the Potential above the

Kinetic Energy, for every later position of the whole system.

A third case is when there is a mixed law of force, such as $r^{-4}-r^{-2}$, so that there is a neutral distance, at which the attraction and repulsion are equal, for all less distances repulsion increasing to infinity, and for greater distances an attraction, rising to a maximum a little beyond the neutral distance, and then slowly decreasing, till it vanishes at an infinite distance. Here, also, it is essential that the Attractive Potential should be reckoned negative, in order that the total of Energy may be constant.

Again, the usual assumption, in modern physical cosmogonies, is that the sidereal systems have reached their present state by condensation from a primitive nebula. Now in a state of rest there is no Kinetic Energy, and if condensation follows, the Potential must be mainly attractive, that is, it will have a negative sign. Or the true statement of the doctrine of Conservation will be, that the sum of the Attractive Potential Energies has a constant excess over the sum of all the Repulsive and all the Kinetic Energy.

The theory of the Conservation of Energy is also subject to these conditions. It fails, and does not hold, if there are any forces in the system, which do not depend on the mutual distance of the particles alone, but vary also with the velocity, or with the time, or with special forms of aggregation, of organization, or types of being. In all such cases the theorem of Dynamics, which has received of late this new title, will no longer apply. Its demonstration cannot be secured by proving that the law of gravitation, and one or two other laws of the same kind, do exist, and have a very wide range of operation.

What is needed is an exhaustive proof that no law of force, except one which makes it depend on local distance alone, can be found in any part of the whole universe.

The maxims involved in the doctrine of the Transformation of Force and Motion, as held by Justice Grove, Mr Spencer, Dr Tyndall, and many other physicists or metaphysicians, appear to be these. First, Motion is Force, or one mode of Force; and Force is Motion, or includes motion as one of its chief modes or forms. Secondly, in the case of impact, the motion of the struck body is the same as the motion of the striking body, only transferred. Thirdly, Kinetic Energy is a present, actual existence. Fourthly, Potential Energy is a present, actual existence. Fifthly, in all cases of dynamical action Potential Energy is transformed into Kinetic, or Kinetic into Potential, but the Energy is the same in either form, and indestructible. Sixthly, the Potential Energy, as defined in the doctrine, is a real Potency. Seventhly, it is the sole Potency, exclusive of any other potencies, which would disturb the equation, and make the total variable. All these maxims, I believe, are untrue, and their error capable of a strict and clear demonstration.

I. The sameness of Force and Motion is the first principle involved in the Conservation of Energy, when viewed as a metaphysical and fundamental truth. This view is implied in many passages of Mr Spencer's work, and among others in these: "If we represent Matter, Motion, and Force, by x, y, and z, we may ascertain x and y in terms of z, but the value of z can never be found" (p. 170). "That which defies suppression in thought is really the force which the motion indicates" (p. 184).

"That which is indestructible in Matter and Motion is the Force they present." "Motion, wherever we can trace its genesis, we find to pre-exist as some other mode of force." "That mode of force, which we distinguish as Heat, is now generally recognized as molecular motion." "The transformation of Heat into Electricity occurs, when dissimilar metals are heated at their point of contact. The transformations of Electricity into other modes of force are still more clearly demonstrable. Each force is transformable directly or indirectly into others, and from the new form or forms it assumes may result either the previous one or any of the rest." In fine, "it is no longer doubted that among the several forms which force assumes quantitative relations are fixed." And the proof assigned for this general principle is the equivalent of a certain increase of Heat, or atomic motion, with a known amount of sensible motion (F. P. pp. 191, 197, 201).

But the same identity is still more distinctly affirmed by Mr Grove in the following passage of his work, 'Correlation and Continuity,' from which Mr Spencer has largely borrowed, and to which he appeals as his main authority.

"Supposing that motion is not an indispensable function of matter, and that it can be at rest, matter at rest would never of itself cease to be at rest. It would not move, unless impelled to such motion by some other moving body, or body that has moved. This proposition applies not only to repulsive motion, but to motion caused by attractions, as magnetism or gravitation. A body at rest would therefore continue so for ever, and a body once in motion would continue so for ever, unless impeded by some other body, or affected by some other force than that which originally impelled it. These propositions may be somewhat arbitrary, and it may be doubted whether they are necessary truths. They have for a long time been received as axioms, and there can be no harm, at all events, in accepting them as postulates."

"The heat which results from friction or percussion is a continuation of the force which was otherwise associated with the moving body, and which when this impinges on another body, ceasing to exist as gross, palpable motion, continues to exist as heat."

Mr Grove seems here to think that he is merely postulating Newton's first law of motion, commonly received as a dynamical axiom. But he is really replacing it by another, which is contradictory to it, and excludes it, and which is irreconcilable with the whole course of reasoning in the Principia from first to last. Mr Grove's maxim is that motion is the only force, and that A moves B, not by the fact of its existence at a definite distance, but by the fact of its motion alone. In one view, for instance, the moon and earth would attract each other with the same force at the same distance, whatever the amount or direction of their motions, or if both were at rest. According to the other, if both are at rest, they can exercise no force on each other, and if they move, the force depends wholly on the velocity and direction of their two motions. The contrast of the two principles is complete. If Mr Grove's is correct, the Principia is one mass of fallacies and false reasonings. If Newton's are real discoveries, the greatest step in advance which Physics have ever made, Mr Grove's remarks are a desertion of the very alphabet of dynamical science. The principle he lays down as an axiom is fatally opposed to the true doctrine of the Conservation of Energy. For this is defined by Sir W. Thomson and Professor Tait as follows:—"If the mutual forces between the parts of a material system *are independent of their velocities,* whether relative to each other, or to any external matter, the system must be dynamically conservative."

"Matter at rest would never cease to be at rest. It would not move, unless impelled to such motion by some other moving body, or body that has moved." This is a statement directly opposed to the law of gravitation, and equally opposed to any probable law of cohesion or ethereal repulsion. Its falsehood is assumed throughout sections VII. and XII. of the First Book of the Principia, which deal with cases of motion beginning, under definite forces, from a state of rest. Indeed it is hard to conceive how any one who makes such a statement can have mastered the simplest definitions and conceptions of dynamical science.

Conservation of Energy, so far as it is a dynamical truth, is based on exactly the opposite principle. By the first law of motion, force is the cause, and motion the effect. Forces may exist without motion in the case of equal and balanced pressures. Motion may exist without force in the ideal case of uniform, unaltered velocity in a straight line. A force *in* the motion, or exercised by the body, simply because it moves, and thus depending on the speed or rate of motion, not the distance, is just what would make the Conservation of Energy untrue, and also has never been detected. All the real discoveries in Physics are of an opposite kind, and all tend to establish and confirm these two principles. Motion is not force, nor any form of force, but simply its effect. Also the forces which do exist, so far as they have been ascertained or conjectured, vary with the distances only, and thus satisfy the condition on which the Conservation of Energy depends.

II. The second maxim, involved in the theory now examined, is the transference of motion, not in a loose and popular, but a strict and proper sense. When one body strikes another at rest, and, the first body ceasing to move, the second moves in its stead, the doctrine teaches that it is the very same motion which has been transferred from the first body to the second. Of course, if one motion were extinguished and a new motion originated in its stead, the whole theory of an indestructible something, varying its forms, but invariable in amount, and firmer than adamant in its essence, must fall completely to the ground.

Let us consider this matter closely. The impact of hard, elastic bodies is the case from which the idea of a simple transfer of the same motion is chiefly borrowed. Let A and B be two equal balls of glass or ivory, moving in the same line with velocities $2a$ and $2b$. After contact or collision, if the elasticity were perfect, they would continue to move in the same line with velocities $2b$ and $2a$. Here we have indeed the semblance of the very same motions continuing, but an amount, $2a - 2b$, being transferred from the first to the second.

What really occurs is different. The motion of the centre of gravity, $a + b$, is not affected by the collision. What is really altered is the relative motion. At first A has a relative velocity $a - b$ towards B and the centre of gravity, and B an equal relative motion towards A. When they meet, both of these motions cease and are extinguished by the mutual repulsion. The two bodies are then relatively at rest, but the adjoining surfaces in a state of compression. These compressions, like the motions, are equal and opposite. The time required by A to destroy the relative motion of B, is the same which B has required to destroy A's relative motion. The two compressions will then generate two new motions, the same in amount as the first, but opposite in direction. Setting aside the common motion, which only obscures the real change, the first motions are AC, BC in opposite directions from A and B towards C, the centre where they meet. The later motions are CA, CB, from the centre backwards, in the direction from which each has come. In the case of a single ball rebounding from a mirror, no one can suppose that the motion has been transferred to another body. Therefore, when two encounter, each retains its own motion, and does not transfer it to the other. The effect is only that the relative part changes its sign. The motion of A, instead of $(a+b)+(a-b)$ becomes $(a+b)-(a-b)=2b$, and that of B, instead of $(a+b)-(a-b)$, becomes $(a+b)+(a-b)=2a$. And the proper statement is, not that so much motion has been transferred from one to the other, but that $2a-2b$ of A's velocity has been destroyed, because its motion was in a direction opposite to the resisting force of B, and that $2a-2b$ has been added to B's velocity, because it was moving in the direction of the impulsive force of A. Part of the motion of A has been abolished. An equal quantity has been added to the motion of B. But still the two motions are not the same, and there is a brief interval between the existence of the first pair of relative motions, and that of the others which replace them, when the balls have separated again.

Motion cannot be of nothing to no place. It must be of something somewhither. It is the change of distance of a real thing, having a definite place with regard to other real things. It may be up or down, north or south, east or west, or partly in each of the three directions. It may be fast or slow, forward or backward. But the motion of

A cannot really be the same motion as that of B. An eastward cannot be the same as a westward motion, a northward as a southward. The motion of a hundred balls eastward, ten feet per second, cannot be the same with the motion of one ball westward, a hundred feet per second. Yet these, and many other paradoxes equally incredible must ensue, if we hold that in collisions the motion is indestructible, and merely changes its domicile, transferring itself from one body to another. The sum total of half the squares of the rates of speed with which every atom or mass in the universe is changing its distance from every other mass or atom, may be a conception of much use in certain dynamical problems. But to claim for it a self-originated, independent existence, due to no Divine Author, and wholly exempt from possibility of change throughout all coming ages is a strange and monstrous inversion of sound reason. The more thoroughly we sift its meaning, the more certainly it will appear to be an unthinkable pseud-idea and self-contradiction.

III. Next, the doctrine assumes Kinetic Energy to be a present actual existence. This is really the half sum of the squares of all the velocities of all the particles or atoms of the universe. It is of course requisite to assume that the number of these atoms is finite. Since there is no natural unit of velocity, an arbitrary unit must be also assumed. The rates of speed, compared with this unit, will be represented by a number, or if the unit be represented by a line, then by other lines. The half-square, in this case, will be a right-angled triangle, and the sum of these areas, for all the pairs of atoms, will be an area representing the total *vis viva* or Kinetic Energy.

Now the motions at any moment do exist, though not as entities, as attributes or relations between the existing particles or atoms. Their sum, also, may be conceived to exist. But in reality the total so measured will be zero; for the centre of gravity of the material universe, viewed as a finite system, must be at rest. In this case the motions up or down, east or west, north or south, must be equal, and neutralize each other. The totals, in the direction of each of three rectangular co-ordinates, must be positive and negative to an equal amount. And thus, when the individuality of each motion is set aside, and they are summed as independent realities, they wholly disappear and annul each other.

But with the half-squares of the velocities the case is different. The square of a negative quantity does not differ from that of a positive, so that all the terms are additive. But is it true that a ball moving twice as fast as another, has actually four times as much motion? Are we at liberty to substitute the half-square of the velocity for the velocity itself, and to treat the total thus found as the sum of the actual motions? The Kinetic Energy would then be the total of present, actual motion, only measured in a somewhat arbitrary way. But in reality the half-square is the integral of the velocity, as the potential function is the integral of the law of force. Hence the more correct view is that the Kinetic Energy is not a sum of the actual motions, but of their integrals; that is, of the velocity summed for all values from perfect rest up to the actual amount. It is thus not the sum of the actual velocities, but of all possible velocities from zero up to the actual relative velocity, for every pair of material atoms. If the whole system were once at rest, and the motions are due to the action of all its internal forces up to any particular time, then it is plain that the Kinetic Energy will be the sum total of all the velocities from that original time of rest up to the actual time. It will not denote an exclusively present existence, but a summation of all the past motions, only limited by some particular moment of time as its termination. On the other hand, if the system were never at rest, but had some original movements, the Kinetic Energy will be an ideal total, in which the motions are summed up as from a state of original rest, when this in fact never occurred. But on either supposition the *vis viva* or Kinetic Energy is not a present, actual amount of motion, but an integral including every past value of each motion down to its real or ideal commencement at some former time.

IV. The Potential Energy, again, if the theory examined were true, must be a present, actual existence. For the Persistence of Force is supposed to mean that a certain fixed amount of it, varying in form continually, and travelling from star to star, or from atom to atom, still abides unaltered in every successive moment. And this requires its actual existence during each moment in succession. But the Potential Energy is not the force acting at the moment, but the integral of that force. In other words, it is the total amount of force that would act through the whole interval of time, till each pair of atoms have travelled from their actual distance to infinity, coalescence, or the neutral distance. In the case of forces purely repulsive the limit is infinity, in those purely attractive it is coalescence. In the case of mixed forces, the limit is the neutral distance, where attraction and repulsion balance each other.

Thus the entire Potential Energy is not, as the theory now examined requires, a force in actual, present existence at any moment whatever. It is a total of possible forces that may hereafter exist, but require for their existence as many distinct periods of possible future time as there are pairs of atoms in the whole universe.

The essence of the doctrine held by Mr Grove, Dr Tyndall, and Mr Spencer, and which the last has made the foundation of his whole theory of Physical Fatalism, is that there is, every moment, an unchanging total of Force, which never varies in amount, while it incessantly changes its form. The Force, then, which persists, must be a present existence. But Potential Energy is nothing of the kind. It is the sum of trillions of trillions of future possibilities of force, ranging through trillions of trillions of different future intervals of time. The element of force, for each pair of atoms, would require a different period for its realization, even if all hindrances were removed, and each pair could exist alone.

The error thus involved in the theory, as a metaphysical axiom, is enormous. The countless intervals of future possible time, in which every pair of atoms, left to their own mutual action, would be able to pass from their actual distance to infinity, zero, or neutrality, are assumed all to coexist, and be included in each passing moment.

V. The fifth datum, assumed in the doctrine, is that Potential and Kinetic Energy are the very same thing, attribute, or substance, its form alone having varied. For the formula in Dynamics does not assert the constancy of either, taken separately, but only of their sum. The one is an integral of force, the other of velocity or motion. But force and motion are not the same. One is the cause, the other the effect. The whole process of continual change depends on this contrast. So also does the whole theory of Dynamics. The first law of motion, the starting-point of Newton's Principia, assumes it. There may be balanced forces, or pressures, without motion. There may be uniform, rectilinear motion, without force. The whole reasoning of dynamical science depends on the clear and sharp contrast between speed or velocity, of which the effect is a uniform change of distance or place, and force, of which the effect is a change in the velocity or speed, or the direction of motion. Thus Potential and Kinetic Energy cannot be the same thing. The integrals of two different things must be different also. Motion is produced by force, and force produces motion. But motion cannot transform itself into force, and force cannot transform itself into motion. The connection indeed is so close, and the relations are so definite, that in loose and popular speech the expressions may be allowed. But in the view of strict science they are always inaccurate.

Forces which balance each other are proved equal by their balancing alone. But accelerating forces can be measured only by the actual acceleration they cause. A total of past force must therefore answer in amount to the total result, if only that result is susceptible of distinct measurement. Now it so happens that, in measuring motion by *vis viva*, or the half-square of the velocity, this condition is fulfilled. And the reason is, that the total *vis viva* is not the sum of present velocities, but of the total velocities, from the initial to the final state of the system. The Past Potential Energy is the total of the causing force throughout that interval, and the Kinetic Energy, or rather its increase, is the effect of that total of force during the same interval. It is natural, then, and inevitable, that the amount of force exercised or expended should answer strictly to the amount of motion generated. Only the two things, though answering in amount, are not the same.

VI. The Potential Energy, to justify the theory, ought to be an actual reality. But it is not even a real potency. It includes the whole force that each pair of atoms would exert in passing from their actual distance to infinity in one case, to coalescence in a second, to the neutral distance in a third.

Now these potencies, to be realized, require each pair of atoms to be successively isolated, and to act on each other undisturbed by any other forces. Taking the case of a trillion of atoms, this would require the fulfilment of a trillion times a trillion of different and irreconcilable conditions. But the atoms cannot isolate themselves. According to Mr Spencer, the annihilation, as well as the creation, of a single particle of matter is inconceivable. But this Potential Energy, which is one of the two main elements in the Persistence of Force, for its mental existence requires us to conceive the destruction of the whole universe, except two atoms, as many times repeated in each single moment, as there are pairs of atoms in all its innumerable worlds.

VII. Seventhly, the doctrine requires us to hold that the Potential Energies it defines are the sole Potentials, and that their variations exactly correspond to the changes of *vis viva*, but with an opposite sign. If there are any other potencies, not included in this reckoning, but with an equal right to be included, of course the proposition that the total is invariable must fail.

Now let us take the simple case of two bodies, A and B, acting on each other by a law of force, which is neutral at the distance c, repulsive for less distances, and attractive for greater. Also let x be any greater, and y any less distance. If the bodies are first at rest at the distance x, they will approach by the attractive force till they reach the distance c, when the potential energy $\phi(x) - \phi(c)$ will be replaced by the *vis viva* $= \frac{1}{2} v^2$. In approaching nearer by this acquired velocity, they will encounter the repulsive force, and come to rest at a distance y, where $\phi(y) = \phi(x)$. The excess of repulsion will then operate to produce a reversed motion, which will be greatest at the neutral distance, and cease at the distance x, where they will be at rest once more.

Here, in passing from the distance x to c, or from y to c, it is plain that the potential $\phi(x) - \phi(c)$, or $\phi(y) - \phi(c)$, has ceased to exist, having done its work. There can be no power left to push from y to c, or to draw from x to c, when the body is at the distance c. But it is just as plain that a new potential has come into being, of the same amount, and opposite in direction. There is a power to lessen motion through the interval from c to x on the one side, and c to y on the other. Under any law of force, depending on the inverse distance, a potency of one sign exists for all less distances, and of the opposite sign for all greater distances. The sum of the two must be always constant. Take the simpler case of a purely attractive force, as the inverse square. Let a body A be attracted towards a centre C, and be at the distance a. Then from a to 0 there is a potential energy to increase the velocity, and from a to infinity, a potential to diminish the velocity. When the body moves from distance a to b, the potential energy to increase velocity over that interval ceases, and is replaced by an equal energy to diminish the velocity in moving from b to a. There are always two potentials, opposite in sign or direction, but equally real possibilities of force, which together fill up the whole range of distances from zero to infinity. With every change of distance, the potential between the two distances changes its direction. A power to accelerate, through a certain interval, is replaced by an equal power to retard in the case of opposite motion, but the total amount of the two potencies is always the same. But the *vis viva*, or Kinetic Energy, does vary indefinitely. And hence the total, when this is included, and all the Potentials are impartially taken into account, will also be variable in the highest degree.

VIII. Force or Motion, on the theory now examined, is invariable in amount, but incessantly changes both place and form. As motion it is transferred from one body to another, and in the same body exists alternately in a Potential or Kinetic form. But on this view no cause whatever is left for this perpetual change. Motion and force having been confounded together, the idea of causality disappears. One part of the mighty whole moves, but can neither push nor pull. Another part pushes or pulls, but cannot move. But some part of the moving portion ceases to move, and begins to push or pull; and some part of that which pushes or pulls ceases to push or pull, and begins to exist as motion. Whence all these changes? Why should energy, which is indifferently force or motion, cease to be force and exist as motion, or cease to be motion and exist as force? The confusion of thought which mingles cause and effect under one ambiguous name, applied in turn to either or to both, leaves the whole series of changes without any possible reason or explanation. What other power compels this blind Titan to occupy a whole eternity with ceaseless and purposeless transmigrations? It is only when force is seen clearly to be distinct from motion, and its cause, that any key to the countless phenomena of the universe can be found. This, accordingly, was the very first step taken by Newton, in those laws or definitions which form the prelude to his immortal discoveries. The first step of the new philosophy is to obliterate this clear line of contrast. And the result is, with all the added facts of modern discovery, to replace a progressive discovery of actual laws of force by a series of high-sounding and ambiguous phrases, which conceal a surprising amount of direct and demonstrable self-contradiction.

If we assume a system, in which a vast amount of motion already exists, and no forces but repulsive ones, acting at a very short distance from the surfaces of bodies, and increasing fast when the distance is diminished, the bodies being perfectly elastic, we shall then have a continual interchange of motions, the collective *vis viva* being always the same, except the part extinguished by collision, and not yet reproduced by the recoil. In this case there is a persistence of Motion or *vis viva*, while of actual force there is no persistence, but a very brief acting in each case of collision. But the case of nature is different. We have a law of attraction, belonging to all distances, and not merely to those which are insensible; and masses are not perfectly elastic, but the ultimate atoms alone. Hence in collision there is only a partial recoil, and the force of compression issues partly in atomic and internal motion. This motion, which we call heat, is so linked with attractive and repulsive forces, that we cannot tell how much is actual motion, and how much, every moment, has been destroyed, and replaced by tension or compression. The first case is that, which the modern advocates of Conservation of Energy, or Persistence of Force, as a grand discovery of science have chiefly in view, and which has moulded their conceptions of all material change. It is really motion, not force, which they look upon as persistent, and the equivalence of degrees of heat with a certain amount of sensible motion is the master-fact to which their appeal is made.

But the real system of forces in nature is widely different. In another stage of theory, an idea arises of a ceaseless dissipation of energy. Thus Sir W. Thomson, one of our highest authorities in mathematical physics, writes as follows:

"It is quite certain that the whole store of energy in the whole solar system has been greater in all past time than at present. But it is conceivable that the rate at which it has been drawn upon and dissipated, whether by solar radiation, or volcanic action in the earth and dark bodies of the system, may have been nearly equable, or even less rapid, in certain periods of the past. But it is far more probable that the secular rate of dissipation has been in some direct proportion to the total amount of energy in store, at any time after the commencement of the present order of things."

With all respect for so distinguished a writer, I conceive that his statement, far from being an undoubted truth, almost exactly reverses the true scientific inference from the laws of nature, so far as already discovered. A ceaseless dissipation of energy is of course irreconcilable with the dictum that invariable sameness of energy is an *à priori* truth. It is true that Mr Spencer, after insisting strongly on this latter doctrine, towards the close of his book adopts the other, its exact reverse, as being also an undoubted conclusion of science. But the statement I have quoted seems to me capable of an easy and simple refutation. Energy, whether Potential or Kinetic, cannot be in constant process of dissipation, unless the matter or ether on which it depends is dissipated also. But the dissipation of matter will increase and not diminish its potential energy. The dissipation of ether, if self-repulsive, would diminish its potential energy. But since in all known cases repulsive forces decrease faster than the attractive, so as to be in excess at small distances, the effect of radiation would be to exchange a certain amount of kinetic and repulsive energy for an equal amount of attractive energy, tending to reverse the process of dissipation. For energy, after all, is an attribute and not a substance, and can never be dissipated without the dissipation of some matter or ether to which it belongs. But, in proportion as these are widely dispersed, the motion will spend itself in overcoming attractive forces, which will be more and more in excess of the repulsive; till a limit is reached, when there must be condensation once more. The doctrine of a ceaseless dissipation of energy is thus, I conceive, wholly groundless, as well as the rival doctrine of its strict and absolute invariability.

A probable view of the atomic forces in actual operation is that they are either self-repulsive, as in the action of ether on ether, or mixed, with a neutral limit, as in the action of matter on matter or on ether. In this case, assuming a system, finite however immense, where even the nearest particles have a distance greater than that of neutrality, and an original state of rest, the later change will be one of condensation, but not indefinite or without limit, with a constant substitution of *vis viva*, or Kinetic Energy, for the Attractive Potential Energy of the first position. And since compression within the neutral distance will be followed by reversed or expansive action, the tendency will be to a growing amount of circular or rotatory motion. There will be, on the whole, no reverse tendency to a later diffusion, but a steady progress from a condition of wider diffusion and absolute rest to one of greater condensation and permanent and steady motion. This agrees with the general conception of the Nebular Theory. But it is wholly opposed to the doctrine of a fixed amount either of Potential Energy or of collective motion, and to the singular hypothesis of a series of alternate evolutions and dissolutions, reaching onward through all eternity.

CHAPTER IX.

ON LAWS OF ATTRACTION AND REPULSION.

THE Law of Universal Gravitation, discovered by Newton, needs to be completed by the discovery of other laws of force, not yet definitely ascertained, before the immense accumulation of facts in every department of Physics can be transformed into a genuine scientific theory, and a solution be found of the great problem of physical change. Many hypotheses have been proposed, either as modest conjectures or more definite theories, from Newton to the present day, and most of them involve the admission of one or more invisible, imponderable fluids or ethers, distinct from common matter. It will be enough to mention some main varieties.

(1) Newton's view. Law of universal attraction, for common matter. Atoms of various shape, finite and unalterable, but endued with active principles, the cause of gravitation and fermentation. An elastic, self-repulsive, discontinuous ether, diffused through all space.

(2) Boscovich's hypothesis. Unextended centres of force, with a law of repulsion for the least distances, and attraction as inverse square for the farthest, and one or three or five intermediate neutral distances.

(3) Young, Fresnel, and Cauchy. Besides matter, a

luminiferous ether, self-repulsive and discontinuous, but its law of relation to matter undefined.

(4) Coulomb and Poisson. Matter and two electric fluids, self-repulsive, and mutually attractive, but their relation to common matter not clearly defined.

(5) Fourier, Poisson. Besides matter, a fluid of caloric, combining with matter by an unknown law, but self-repulsive, with special laws of conductivity and equal diffusion.

(6) Mosotti. Matter and one electric fluid. Atoms of matter spherical and self-repulsive. Ether or electric fluid, discontinuous and self-repulsive. Matter and fluid mutually attractive. All the three laws are of the inverse square. Three other subsidiary laws are assumed, inconsistent with each other.

(7) Exley. Matter, unextended centres of force, attracting by the law of the inverse square, but the force changes sign, and becomes repulsive at a neutral distance, different in different atoms. Chemical atoms distinguished by different constants of force, and different neutral distances. Two kinds of ether, luminiferous and electric, each consisting of unextended centres of force and self-repulsive.

(8) Norton. Matter, spherical finite atoms, mutually attractive. Electric ether, self-repulsive, attracting matter, and attracted by it. Luminiferous ether, more subtile than the other, but also self-repulsive, and attracted by common matter.

(9) Challis. Matter, spherical, finite atoms, devoid of force, around which ether is condensed. Ether, a plenum and continuous, but of variable density, and its fundamental law, pressure varying as the density. Attraction and all other phenomena results of this varying pressure.

(10) Bayma. Matter of two kinds, positive and negative. The first attracts, the second repels, all atoms, and by the law of the inverse square. Both are unextended centres of force.

(11) Helmholtz. Atoms, vortices of a revolving fluid; supposed to be permanent, when once formed, by an assumed law of continuity.

(12) My own hypothesis, in "Matter and Ether." Matter, unextended atoms, as Boscovich, but attracting by simple law of the inverse square, as Newton. Ether, unextended monads, attracting matter by a higher inverse power than the second, and self-repulsive by a still higher inverse power. Each atom of matter inseparably combined with one of ether, so as to have a neutral distance. Chemical atoms compound, being the first step in the composition of these primary atoms.

(13) Grove, Brooke, Winslow. Matter, minute, finite atoms, endowed both with attractive and repulsive forces, their laws undefined. No ether distinct from matter.

All these views, except the last, agree in offering an hypothesis more or less definite, and capable of becoming the subject of mathematical reasoning and calculation. Some of them, I think, involve a secret inconsistency, and others are needlessly complex in their assumptions. But at least they fulfil the first condition of a physical theory, and admit of being theoretically unfolded, so that the results of this development may be compared with those of actual experiment.

The doctrine laid down in the "First Principles" has a character precisely opposite. It is a physical theory composed simply of abstract, metaphysical terms, that may be applied indifferently to a thousand varying hypotheses, and cannot therefore advance us a single step in the path of genuine discovery. But it has a still worse fault. It is not only vague and wholly indefinite, but self-contradictory. The first elements of the problem seem never to have been distinctly apprehended, so that, instead of travelling beyond Newton's great discovery to further triumphs, we are led backward into a region of mere nebulosity and confusion.

The two following passages, which I give at length, from the vital importance of the subject to the whole issue before us, contain the Physical Creed in the First Principles, which has to serve as the only substitute for all religious creeds whatever. §§ 18, 74, pp. 60, 223:

"Light, Heat, Gravitation, and all central forces, vary inversely as the squares of the distances, and physicists in their investigations assume that the units of matter act upon each other according to the same law; an assumption which indeed they are obliged to make, since this law is not simply an empirical one, but one deducible mathematically from the relations of space, one of which the negation is inconceivable. But now, in any mass of matter which is in internal equilibrium, what must follow? The attractions and repulsions of the constituent atoms are balanced. Being balanced, the atoms remain at their present distances; and the mass neither expands nor contracts. But if the forces with which two adjacent atoms attract and repel each other, both vary inversely as the square of the distance, AS THEY MUST; and if they are in equilibrium at their present distances, as they are, then necessarily they will be in equilibrium at all other distances. Let the atoms be twice as far apart, and their attractions and repulsions will both be reduced to one-fourth of their present amounts. Let them be brought within half the distance, and both will be quadrupled.

Whence it follows that this matter will as readily as not assume any other density, and can offer no resistance to any external agents. Thus we are obliged to say that these antagonist molecular forces do not both vary inversely as the square of the distance, WHICH IS UNTHINKABLE; or else that matter does not possess that attribute of resistance, by which alone we distinguish it from space, which is absurd. While, then, it is impossible to form any idea of Force in itself, it is equally impossible to comprehend either its mode of exercise or its law of variation."

"The Absolute Cause of changes, no matter what may be their special natures, is not less incomprehensible in respect of the unity or duality of its action than in all other respects. We cannot decide between the alternative suppositions, that phenomena are due to the variously conditioned workings of single force, and that they are due to the conflict of two forces. Whether, as some contend, everything is explicable on the hypothesis of universal pressure, whence what we call tension results from inequalities of pressure in opposite directions; or whether, as might be with equal propriety contended, things are to be explained on the hypothesis of universal tension, of which pressure is a differential result; or whether, as most physicists hold, pressure and tension everywhere coexist, are questions which it is impossible to settle. Each of these three suppositions makes the facts comprehensible, only by postulating an inconceivability. To assume a universal pressure confessedly requires us to assume an infinite plenum, an unlimited space full of something which is everywhere pressed by something beyond; and this assumption cannot be mentally realized. That universal tension is the immediate agency to which phenomena are due, is an idea open to a parallel and equally fatal objection. And however verbally intelligible the proposition that pressure and tension everywhere coexist, yet we cannot truly represent to ourselves one ultimate unit of matter drawing another, while resisting it.

"Nevertheless this last belief is one which we are compelled to entertain. Matter cannot be conceived, except as manifesting forces of attraction and repulsion. Body is distinguished in our consciousness from Space by its opposition to our muscular energies; and this opposition we feel under the twofold form of a cohesion that hinders our efforts to rend, and a resistance that hinders our efforts to compress. Without resistance there can be merely empty extension. Without cohesion there can be no resistance....We are obliged to think of all objects as made up of parts that attract and repel each other, since this is the form of our experience of all objects.

"By a higher abstraction results the conception of attractive and repulsive forces pervading space. We cannot dissociate force from occupied extension, or occupied extension from force; because we have never an immediate consciousness of one in the absence of the other. Nevertheless we have abundant proof that force is exercised through what appears to our senses as a vacuity. Mentally to represent this exercise, we are obliged to fill the apparent vacuity with a species of matter, an ethereal medium. The constitution we assign to this medium, however, like the constitution we assign to solid substance, is necessarily an abstract of the impressions received from tangible bodies. The opposition to pressure which a tangible body offers to us is not shewn in one direction only, but in all directions, and so likewise is its tenacity. Hence the constitution of those ultimate units through the instrumentality of which phenomena are interpreted. Be they atoms of ponderable matter or molecules of ether, the properties we conceive them to possess are nothing but these perceptible properties idealized. Centres of force attracting and repelling each other in all directions are simply insensible portions of matter, having the endowments common to sensible portions, endowments of which we cannot by any mental effort divest them. To interpret manifestations of force which cannot be tactically experienced we use the terms of thought supplied by our tactical experiences, and for the sufficient reason that we must use these or none.

"It needs scarcely be said that these universally coexistent forces of attraction and repulsion must not be taken as realities, but as our symbols of the reality. They are the forms under which the workings of the Unknowable are cognizable by us, modes of the unconditioned as presented under the conditions of our consciousness. But while knowing that the ideas thus generated in us are not absolutely true, we may unreservedly surrender ourselves to them as relatively true, and may proceed to evolve a series of inferences having a like relative truth."

The following are the maxims taught in the previous passages, and made the basis of the whole philosophy of Physical Fatalism. (1) That gravitation varies by the law of the inverse square of the distance. (2) That all other central forces vary by the same law. (3) That physicists are obliged to assume this law, because it results from the mathematical properties of space. (4) That any other law of variation is inconceivable. (5) That all matter repels as well as attracts by this same law. (6) That this is an *à priori* truth, and its denial unthinkable, though it leads to an evident absurdity. (7) That consequently of Force, its mode of exercise, or its law of variation, nothing can be really known. It is itself the Unknowable. (8) That we cannot decide whether pressure alone, that is, repulsion, or tension alone, that is, attraction, is the real cause of phenomena, or both combined. (9) That for one atom to attract and repel another at the same instant, is inconceivable. (10) That though inconceivable, it is the belief we are compelled to entertain. (11) That all the parts of a solid body do both attract and repel each other at the same instant. (12) That our conception of an ethereal medium is necessarily the same. (13) That our conception of centres of force is that of small coherent masses, resisting compression. (14) That these coexistent forces of attraction and repulsion, after all, are not realities, but symbols of some reality. (15) That the Unknowable is cognizable under these forms. (16) Further, that while the unconditioned must lie beyond our consciousness, since to think is to condition; and beyond our knowledge, since it is the unknowable, it has modes which lie within our consciousness, and are cognizable. (17) Lastly, that conclusions drawn from an assumption, the truth or falsehood of which is wholly uncertain, may be and should be accepted unreservedly as relatively true, though they are nothing more than symbols of some unknown and inconceivable reality.

The first of these maxims is Newton's great discovery, and the true starting point of all later advances in physical science. The ninth also is a self-evident truth, which proves that the fifth, sixth, tenth, eleventh and twelfth are direct self-contradictions. All the others are untrue, and some of them self-contradictory and absurd. Let us examine them in order.

First, all central forces, besides gravity, do not follow

the simple law of varying as the inverse square of the distance. In this case there could be no such thing as repulsion in the universe, supposing attraction to exist. For the repulsion, if greater, equal, or less at one distance, would then be greater, equal, or less at all other distances. If greater, it would extinguish the attraction, and leave a surplus of repulsion only. If equal, there would really be neither repulsion nor attraction, but a state of perpetual rest with no force whatever. If less, then there would be always a mutual attraction, and no repulsion. But it is plain that, besides gravitation, there is a cohesive force, which increases faster when the distance is lessened, and a repulsive force, which increases faster still, so as to result always in resistance to ultimate compression.

Secondly, physicists are not obliged to assume this law of the inverse square as the sole law of force. In fact, they have not so assumed, but just the reverse. Newton, the foremost of them, plainly assumes many other laws to be possible, and a large part of the Principia is employed in tracing out their various consequences. In his Optics, also, he plainly assumes that both matter and the ethereal fluid have other laws or 'active principles,' distinct from the law of universal gravitation. The theory of Boscovich, again, is wholly based on the assumption of one or more laws of force, of a transcendental kind, so that the curve of force would cross the axis, one, three, or five times. The hypotheses of Cauchy on Ether, and of Poisson on Caloric and Capillary Attraction, and one half at least of other dynamical theories, assume some law or other, varying more rapidly than the inverse square. Thus the facts are just the reverse of what Mr Spencer has here assumed.

Again, there is nothing in the properties of space to compel the assumption of this law, to the exclusion of every other. They can lead to the Newtonian law only by the help of two postulates; first, that the total force exercised by an atom at any one distance is the same as at any other; and next, that space is a plenum of uniform density. Now the first of these is so far from being a self-evident principle, as to be a direct reversal of our usual and early impressions, since all action of one body on another seems confined to insensible distances. And the second is so far from being self-evident, that it is certainly untrue. Most physical philosophers agree with Lucretius and Newton in rejecting a continuous plenum, as rendering all motion impossible. And all those who hold it deny that its density is invariable.

Thirdly, that matter both attracts and repels by the law of the inverse square is no necessary truth. On the contrary, it is a plain self-contradiction, a notion due only to confusion of thought, and strictly unthinkable. It is astonishing that Mr Spencer, who notices the absurd conclusion to which it leads, should not have been deterred from propounding it, in defiance alike of history and common sense, as an inference from the geometrical laws of space, which all physicists have been compelled to assume. Repulsion by a higher law than the inverse square leads to results in harmony with known experience, a neutral distance, repulsion within that limit, and attraction beyond. But if the law of variation were the same, the less would have no effect but to diminish the greater, and either no force would be left at any distance, or one force only, whether of attraction or repulsion, at all distances great or small. In the first case, Unknowable Force, the new divinity, becomes a cipher, and is self-extinguished. In the second case, we shall have either an incoherent multitude of self-repulsive atoms, all receding from each other, and losing themselves in empty space; or an equally incoherent multitude of atoms, confusedly condensing more and more, and whirling round each other with infinite turmoil, without solidity or cohesion.

Fourthly, it is wholly untrue that nothing can be known of Force, and the laws of its variation. If it were true, it would condemn Physics, as a science, to the same grave to which Mr Spencer has consigned Theology. But the exact reverse is true, as is plain from Newton's great discovery alone. And the wonderful thing is that, in the same paragraph, this main result of profound mathematical reasoning, compared with the facts of astronomy, should be promoted into an *à priori* truth, which every one must have known without Newton's help, because its reverse is unthinkable, and set down in the category of things impossible to be known at all.

We seem now standing on the verge of further discoveries of laws of force, and their variation, to complete and perfect the great work which Newton, in the Principia, has so effectually begun. But this progress can never be made by mistaking direct and open self-contradictions for *à priori* and self-evident truths, and laying them down for the basis of a new and improved scheme of physical philosophy.

Fifthly, it is quite easy to decide that both attractive and repulsive forces do exist, and not attraction only or repulsion only. They cannot, however, coexist for the same pair of atoms at the same distance, in any system where all the forces are functions of the distance alone. Now this last is the condition required in order that Conservation of Energy may be an actual truth.

"Matter cannot be conceived, except as manifesting forces of attraction and repulsion. We are obliged to think of all objects as made up of parts that attract and repel each other."

It is a constant feature of Mr Spencer's philosophy that a statement and its contradictory are repeatedly affirmed with equal positiveness, occasionally in adjoining pages, or the same paragraph; while sometimes both of these contradictions are promoted into the dignity of self-evident truths. Here we have been told, just before, that we cannot decide whether the phenomena of change arise from both attractions and repulsions, or from one of these two kinds of force only. And now we are told the exact reverse, that we are obliged to believe in that duality of the action of force, which has been just pronounced to lie beyond the range of our knowledge, and to be inconceivable.

This constant oscillation and confusion of thought is most wearisome and vexatious for any reader, who desires really to gain insight into the questions in debate. Here the truth will be found in the following remarks. We cannot conceive a pair of atoms both repelling and attracting each other at the same moment. One or other force must be in excess, or there will be no action whatever. Yet we can conceive of the same point combining a double action, attractive and repulsive, each varying by a different law. In this case there will be a neutral distance where the two are equal, and their result zero, and there will be excess of repulsion on the one side, and of attraction on the other. We may thus speak of them as attracting and repelling equally at the neutral distance, though really there is no action at all, neither push nor pull, and the difference of the two values is in every case the sole acting force. Cohesion and resistance to compression, with a neutral intermediate position, will all result naturally from the union of repulsive and attractive force in the same centre, the repulsion following a higher law of variation than that of gravity or cohesive affinity. Instead of this self-consistent view, Mr Spencer's theory is made up of four assertions, all separately untrue, and also inconsistent with each other: that the laws of the variation of force are wholly incognizable; that it varies as the inverse square of the distance by a geometrical necessity, and no other law of variation is conceivable; that attraction and repulsion vary therefore by the same law; and still that both attraction and repulsion do coexist, which, if the law be the same, is manifestly impossible and absurd.

Sixthly, it is not true that our conception of an ethereal medium, and of solid matter or gravitating substance, is the same. Different views may be taken of the exact nature and constitution of the luminiferous ether. Nearly all, however, who affirm its existence, will agree that it is self-repulsive. This view is common to Newton, Young, Fresnel, Herschel, Cauchy, Challis, Stokes, and almost all other writers on physical optics. That ether has "inertia, but not gravity," is made by Herschel one fundamental principle of the theory. On the other hand, mutual attraction is now almost the foremost element in the definition of common matter. Constancy in the amount of matter is thus usually inferred from constancy of weight, and from that alone. But the two conceptions, of centres which are self-repulsive only, and of others which are repulsive at insensible distances and attractive at all sensible distances, are different and opposite on the very point in which their identity is here affirmed.

Seventhly, the conception of a force-centre is not the

same as that of a small coherent mass, resisting compression. For both coherence of parts and resistance to compression imply manifoldness and plurality, while a force-centre is a unit, and excludes the notion of plurality. Many force-centres may and will cohere, when brought near together, approaching till they reach the neutral distance. They will also resist the compression which seeks to bring them still closer together. But both of these characters involve the presence of a considerable number of such units, and are wholly inapplicable to one such force-centre, taken alone.

Eighthly, it is true that attractive and repulsive forces must be measured by the velocities they generate in a unit of time, and these velocities, again, by the space traversed in a unit of time, nor can we easily conceive of their being measured in any other way. But we do not conceive them as symbols of unknown realities. They are mysterious realities themselves. They are defining attributes or characters of all material substance. Place and force combined, so far as we can discern, are the essence of matter. Motion is an accident, which belongs alike to masses and to units or atoms. It is not an essential quality, for matter will still be matter, though at rest. Extension, cohesion, compressibility, and elastic recoil, or resistance to compression, are attributes, not of the units or atoms, but of material masses or bodies, composed of a multitude of such units, closely united together. There is no resemblance whatever between a treatise on dynamics, and one on poetical or ecclesiastical symbolism. And indeed there can be no symbolism, that deserves the name, of an object or series of objects utterly unknown.

The two latest maxims of the theory simply reproduce errors, which have been refuted before. But it may be well to point out once again the fundamental misconception, to which we owe the incessant oscillation and confusion of thought, which marks the whole reasoning in this doctrine of the Unknowable.

In the study of nature and the universe, three things need to be carefully distinguished. The first is the phenomena, momentary, fleeting, evanescent, of which one perishes before another is born, so that almost an infinite number of them need to be combined, in order to attain any knowledge of things or persons, or conclusion of settled science. The next are the causes of phenomena, things and persons, true causes, which abide and endure, while the phenomena are born and perish; but secondary and limited, whether by place, as in material objects, or by consciousness, memory, and reflexion, as in living persons. The third is the Great First Cause, apart from whom these second causes cannot be explained, the True, Supreme, Absolute, Infinite Being. The first, phenomena, we experience in successive moments, but cannot properly be said to know. A vague, dim memory is all that we can attain. But second causes, persons and things, since they are permanent, and defined by place or consciousness, admit of being known, and thus constitute the sciences of Physics and of Humanity. The First Cause also may be known, though this knowledge is higher, nobler, more mysterious, and harder to attain; and this knowledge constitutes the science of Theology.

The doctrine of the Unknowable, and its equivalent, the Philosophy of the Unconditioned and the Conditioned, first of all confounds mystery with self-contradiction. It multiplies antinomies, or pairs of contradictories, and pronounces each alternative alike inconceivable and unthinkable. Thus a truth too large and vast to be fully comprehended, or seen on all sides, though we may have the fullest evidence of its truth, is placed on the same level with a contradiction and an absurdity.

In the next place, Second Causes and the First Cause are confounded together. The Pantheism, disowned in words, is adopted fully in substance. For its essence is not the doctrine that the universe is self-created, but that God is simply the sum total of all things. Thus things, persons, and any higher power, the cause of both, are all blended together under one common title, the Unknowable. All phenomena, which alone are knowable, are faint or vivid manifestations of this unknowable, Pantheistic Being, the Deity which is the Universe.

Here, then, the doctrine encounters an insuperable difficulty. Its first result will be that phenomena alone are knowable, and that all causes, matter, motion, force, conscious living persons, and the Absolute, the Infinite Being, the First Cause of all things, are one and the same, the Great Unknown and Unknowable. But phenomena do not admit of knowledge. Each sensation can only be felt while it lasts, and remembered imperfectly, when it has ceased to exist. So we have to invent some middle terms, by which the perishable phenomena may gain some element of permanence, and the Unknowable admit, more or less, of being known. And this end is sought to be attained by the two phrases, relative realities, and modes of the Unknowable. Second causes, things and persons, are alternately confounded with the Absolute Reality, when they share in its character of being utterly inscrutable and unknowable; and with Phenomena, when they are symbols and shadows only, and still the subjects of a genuine knowledge. But unfortunately all the actual discoveries of science have been in the region, not of mere phenomena, but of the causes of those phenomena; that is, material substances, endued with attractive and repulsive forces; and minds, living persons, endued with the gifts of sentience, memory, thought, reason and will. And hence the philosophy has to treat these second causes, and their fundamental laws, alternately as unintelligible, unthinkable mysteries or absurdities, and as self-evident, trustworthy, uncreated, eternal, *à priori* truths.

Let us now resume the subject from the beginning, and unfold the axioms of genuine Physics, which are in full harmony with those of Morals and Theology, in contrast with the nebulous confusion of Religious Nihilism and Physical Fatalism.

Science starts from the experience of phenomena. They are the data which it has first to register, to combine, to analyze, and to re-combine, in order to determine the causes on which they depend, and thereby from past experience to anticipate future changes. But phenomena, as such, are not knowable. The momentary sensation, ending as soon as felt, does not deserve the name of knowledge. For knowledge must be of some thing that abides and endures. What perishes, while we are thinking of it, or before we reflect upon it, cannot be really known.

Our first step must be to rise from effects, which are transient and momentary, to their permanent causes. Perceptions and reflexions are the two main classes of phenomena. And these, when joined with the idea of causation, suggest causes of two kinds. The first kind is Matter, or outward, real, material objects. The second species of cause is Mind; first and directly, our own Mind, the self which persists and is the same in every act of self-consciousness, and next, indirectly, the Minds or Living Persons of our fellow-men, found by experience to have the like capacities of thought and feeling with ourselves.

But Mind and Matter, though causes in contrast to fleeting phenomena, have not the characters of self-existence, or of the final and supreme reality. Matter is local, multiple, manifold, and variable, and lower in dignity than Mind itself. Mind also, in ourselves and our fellow-men, is linked with Matter, limited, dependent, and conscious of moral weakness and imperfection. A like process of thought to that which transfers us from momentary phenomena to permanent realities, things and persons, when repeated, raises us from things and persons to the conception and knowledge of the Supreme Reality, the First Cause and Final End, of the whole created universe.

The first part of these two stages of thought transfers us from the phenomenal to the real, from the ever-shifting phantasmagoria of mere sensations, a cloud-land of fog and mist, to the high table-land of definite science, where both Physics and Humanity find their proper dwelling-place and home. The second raises us still higher, from the table-land of natural and human science to the mountain tops of Theology. It plants our feet on the floor of the heavenly temple, enables us to know "Him who is from the beginning," to see Him who is invisible, to gaze upon the moral perfections of the Infinitely Good and Wise, and to see light in the light of heaven.

Space, Time, Matter, Motion, Force, the ultimate ideas in Physics, are all mysterious. For we know in part, and wherever there is only partial knowledge, what remains unknown constitutes a mystery still unsolved. But mystery is not self-contradiction or absurdity. Partial knowledge is still real knowledge, and this real knowledge is capable of growth and extension. The landscape which lies within the horizon may be clearly seen. The horizon itself may recede, and the visible landscape be enlarged. But still there is an horizon, and what lies beyond it is a mystery. To confound this presence of a felt mystery with absurdity and contradiction is fatal to all genuine science. The Ultimate Scientific ideas are mysterious, and still there is a true and genuine science of Physics. The Ultimate Religious Ideas are mysterious, and still Theology, no less than Physics, is a genuine science. The ascent is higher, but the pathway is the same. The progress which leads from the mists in the low valley to the mountain side, if we pursue it further, will bring us to those summits of thought, where we can gaze on the Supreme and Perfect Goodness, and have a vision of the things unseen and eternal, on which all the changes in the lower world of men and nature depend.

Matter is real, and Mind is real. Because they are real, and not phenomenal, they may be known, and become the subjects of genuine science. Matter is localized Force, or dynamized Position. As force, it is contrasted with mere empty space. As localized force, it is contrasted with force of a higher kind, with the energies of thought in minds or living persons, and with almighty power in the Supreme and Infinite Being. As composed of units or monads, it gives birth to one complete science, that of Number, or Arithmetic. As local, it gives birth to a second science, that of Geometry. As force, but localized, it gives birth to a third science, that of Physical Dynamics.

Phenomena, as such, cannot be known. Sensations can only be experienced for the moment, and expire in their birth. But through phenomena and sensations we gain a knowledge of their causes, the real things, outside of us, and in space, on which these appearances and sensations, varying every instant, depend. These causes alone are the objects of true science. And the Cause of these causes, the great First Cause, is the object of the highest of all sciences, Christian Theology. Other sciences minister to the wants of man in this present life. They enrich it with manifold inventions of art, promotive of social comfort, and with a large variety of intellectual occupations, by which the reasoning faculties are strengthened, and man's power over nature is increased. But to the deeper questions of the human conscience they give no answer, and for the worst ills of human life they provide no remedy. Those questions can only be answered, the cure for those ills provided, by knowledge of a higher and nobler kind, which deals with the relations between a holy and wise Creator, and the creatures whom He has made. The benefits which physical science has to bestow on mankind are great and manifold, and still are liable to be overbalanced by greater evils, when the growth of knowledge leads to intellectual pride, and pride breeds open profaneness and contempt and rejection of all higher truth. But all truth is naturally allied. And of all sciences the noblest and the most profitable is that which reveals to us the character, the purposes, and the perfections, of Him who is the Fountain of all being, the supremely good and wise. "For this is life eternal, to know thee, the Only True God, and Jesus Christ, whom thou hast sent."

CHAPTER X.

ON CHOICE AND WILL IN PHYSICAL LAWS.

PHYSICAL Fatalism, as a complete theory, reduces all the changes of the universe to complex and varied forms of atomic motion. It excludes all ideas of freedom, choice and will, even in the actions of all men. But it also requires the resolution of physical laws themselves into the result of some blind necessity. They must be original and self-created, and capable of being explained without any reference to the choice of a wise Law-giver and supreme Intelligence.

The acceptance of this negative creed has been laid down, by some of its advocates, as the test of a really philosophic mind. Mr Spencer, in the first and second editions of his First Principles, applied it expressly to Newton's law of gravitation. Physicists were obliged to assume this law, because it results from the necessary conditions of geometrical space. But in his third edition, after fifteen years, the statement is silently withdrawn. Its historical falsehood, if not its theoretical absurdity, seems at last to have been detected by its author. But no explanation is offered, and no open retractation is made. Yet it is plain that the affirmation, however untrue in point of fact, and self-contradictory in theory, is closely linked with the whole scheme of philosophy, in the forefront of which it stood so long. If physical laws can exist only by the free choice of some wise Lawgiver, the doctrine of a Moral Governor is established. The walls of the dark prison-house of necessity are thrown down, and we may breathe freely once more.

Genuine Philosophy, according to M. Comte, is employed in the grouping of phenomena, and in that alone. It excludes alike all Theology, or the acknowledgment of a Moral Governor, the great First Cause, and all Metaphysics, under which we are to place the ideas of cause and effect, as well as substance and vital power. But his disciples or successors have speedily diverged into an opposite view. The metaphysical ideas he denounced are brought in anew with great pomp and state, and made the main feature of their improved philosophy. Instead of excluding the idea of Force, as metaphysical, they seem ready to invest it with Divine attributes, and place it on the throne of the universe. It is a Something uncreated, invariable, indestructible, almighty and eternal.

The doctrine of materialism, in its recent form, looks upon force, self-acting by pure necessity, as the main subject of all philosophy. It has been thus expressed in the Lay Sermons. "It is no less certain that the existing world lay potentially in the cosmic vapour, and that a sufficient intelligence could, from a knowledge of the properties of the molecules of that vapour, have predicted, say the state of the fauna of Britain in 1869 with as much certainty as one can say what will happen to the vapour of the breath on a cold winter's day." And again, "As surely as every future grows out of present and past, so surely will the physiology of the future extend the realm of matter and law, until it is coextensive with knowledge, with feeling, and with action."

In Dr Tyndall's article on "Miracles and Special Providences" the same doctrine is advanced in its most definite and articulate form. He writes as follows:

"Kepler had deduced his laws from observations. As far back as those observations extended, the planetary motions had obeyed those laws, and neither Kepler nor Newton entertained a doubt as to their continuing to obey them. Year after year as the ages rolled, they believed that those laws would continue to illustrate themselves in the heavens. But this was not sufficient. *The scientific mind can find no repose in the mere registration of sequence in nature.* The further question intrudes with resistless might—Whence comes the sequence? What is it that binds the consequent with the antecedent in nature? The truly scientific intellect never can attain rest, until it reaches the *forces* by which the observed sequence is produced. It was thus with Torricelli, it was thus with Newton; it is thus preeminently with the truly scientific man of to-day. In common with the most ignorant, he believes that spring will succeed winter, that summer will succeed spring. But he knows still further, and this knowledge is essential to his intellectual repose, that this connexion, besides being permanent, is under the circumstances *necessary*, that the gravitating force *must* produce the observed succession of the seasons. Not until this relation between the forces and the phenomena has been established, is the law of reason rendered concentric with the law of nature, and not until this is effected does the mind of the scientific philosopher rest in peace....If the force be permanent, the phenomena are *necessary*, whether they do or do not resemble anything that has gone before."

Here the principle of M. Comte is entirely reversed.

The mere registration of sequences is pronounced incapable of satisfying the scientific mind. It cannot rest, until it attains a knowledge of the forces, and their laws of action, on which the sequence must depend. So far the remark is perfectly true. But having thus deserted and reversed one main article in the original creed of Positivism, the writer departs just as widely from the truth on the other side, and makes it the essence of the scientific instinct to look on physical laws as acting of themselves, by some inherent necessity alone.

The authority of Newton is invoked to confirm this new creed of science, and reprove the ignorance of believers in miracles and special providences, who are called blind leaders of the blind. But no attempt to solve the deepest and noblest problems of the natural and moral universe can succeed, which starts by reversing the plainest facts in the history of science. The maxim advanced under the alleged sanction of Newton's name is the very same, which Newton himself rejected and condemned, as utterly opposed to sound reason. He writes of it as follows:

"This most beautiful system of the Sun, Planets and Comets, could only proceed from the counsel and dominion of an intelligent and powerful Being. And if the fixed stars are the centres of like systems, these, being formed by the like wise counsel, must be all subject to the dominion of One; especially since the light of the fixed stars is of the same nature with the light of the sun, and from every system the light passes into other systems....

"This Being governs all things, not as the soul of the world, but as Lord over all. The Supreme God is a Being eternal, infinite, absolutely perfect. We know Him only by his most wise and excellent contrivances of things, and final causes; we adore Him for his perfection; but we reverence and adore Him on account of his dominion. For we adore Him as his servants; and a God without dominion, providence, final causes, is nothing else but Fate and Nature. Blind metaphysical necessity, which is certainly the same always and everywhere, could produce no variety of things. All that diversity of things which we find, suited to different times and places, could arise from nothing but the ideas and will of a Being necessarily existing. And thus much concerning God, to discourse of whom from the appearances of things does certainly belong to Natural Philosophy."

At the close of the Optics, Newton's much later work, another statement of the same kind recurs once more, and ends with these significant words: "If Natural Philosophy in all its parts shall at length be perfected, the bounds of Moral Philosophy will also be enlarged. For so far as we can know by Natural Philosophy what is the First Cause, what power He has over us, and what benefits we receive from Him, so far our duty towards Him, as well as that towards one another, will appear to us by the light of nature."

Modern Fatalism, then, and the creed of Newton, the foremost of all physical discoverers, are not the same, but diametrically opposed. The maxim of a necessity inhering in all physical laws, for which Dr Tyndall would make him sponsor, is the one doctrine which, at the close of both his immortal works, he strongly denounces and condemns. But since even his authority is lightly esteemed by some of the aspiring advocates of the new philosophy when it is found adverse to them, it may be well, apart from names, to examine the question in the light of reason alone. Are Physical Laws the results of some inherent necessity, which shuts out every alternative law, and makes it impossible and unthinkable? Or do they exist side by side with many possible alternatives, equally conceivable? Is not their discovery, for this very reason, the work of patient induction, which compares observed facts with the reasoned results of different possible laws, and thus decides which of them really exists? It must then be clear that they are not necessary. Their existence can only be explained by the choice and wise counsel of some Divine Lawgiver, the Lord and Maker of the whole universe.

Let us begin with the Newtonian law itself. "Every particle of matter attracts every other particle and is attracted by it, with a force which varies inversely as the square of their mutual distance." Is this law necessary, like the truth that a whole is greater than its part, or that two and two are four, or that every right-lined three-sided figure must have three angles? Or is it one out of many conceivable laws, which are possible in themselves, so that it must be what it is only by the will and choice of a supreme Lawgiver?

The answer is quite plain. There can be no *à priori* necessity that every particle should act on every other at all at every distance, and in any position. Indeed Mr Spencer, after affirming for fifteen years that physicists have adopted and accepted this law of the inverse square, because any other is unthinkable, has replaced it suddenly by the opposite assertion, that action at a distance, by any rule of variation whatever, is "positively unthinkable," and that action equal in amount, whether the intervening space is empty or occupied, is equally incomprehensible and inconceivable. He gives no word to explain this abrupt transition, by which that is an inconceivable absurdity to-day, which yesterday was proclaimed a necessary and *à priori* truth.

Again, if this law were necessary, no other law of force could exist. It must reign alone. There could then be no such thing as repulsive force anywhere in the universe. This flatly contradicts facts still more numerous and familiar than those on which the Newtonian law itself depends.

In the third place, if there be only one law, extending to all particles and all distances, repulsion and attraction would be equally possible in their own nature. Motion is always equally conceivable in two opposite directions. And since force is simply the cause of motion, where an attractive force is conceivable, a repulsive one must be so likewise. If atoms must either push or pull, there can be no inherent reason why they do one and not the other.

Fourthly, there can be no inherent necessity why each particle should act on all others, and not on some only, and at all distances, not some only. This wide and universal extent of the law, far from being self-evident, requires a great mental effort to receive. The proof in the Principia is inductive and cumulative, through five or six successive stages, for Jupiter and its satellites, the Sun and its planets, the Earth and its moon, and the component parts of the earth. It is clear, at every step, that a more limited law is quite conceivable, and that its extension to sun, planets, comets, satellites, fixed stars, and their component portions, can be no abstract necessity, but a proved result of patient induction alone.

Fifthly, this law of the inverse square is only one out of a great number, equally conceivable. Those of the inverse distance, the inverse cube, fourth, fifth, and sixth power, are not merely conceivable, but have been analysed in works of Dynamics, and the nature of the curves or orbits inferred by strict reasoning. No small part of the Principia is occupied with deductions of this very kind.

No doubt the law of gravitation, as the inverse square, might be conceived to result, as in the case of the diffusion of light and heat, from the transmission of a central disturbance, definite in amount, which spreads itself equally over the spherical surfaces concentric to each other. But then, in replacing a direct law of force of a very simple kind by the laws of geometry as to spherical surfaces, we need to introduce some other positive laws, and conditions far more complex than the one we strive to explain and supersede, and more widely remote from any appearance of necessary truth. We require, for instance, a medium of unknown and complex constitution, a speed of transmission far greater than that of light, and needing thus a second and still more elastic ether, a force resulting from waves of disturbance, or differences of pressure, and yet a perfectly equal density in all places, and for all concentric surfaces. These conditions, I believe, are really incompatible. And even if they could be reconciled, they are far more complex than the law they are intended to replace, and involve a far larger amount of non-necessary elements.

Lastly, the law of Conservation of Energy, so highly celebrated by modern fatalists, and made the main pedestal of their philosophy, is diametrically at variance with their chief doctrine, or the necessity of physical laws, and especially the best known of them all, the law of gravitation. For what is its real meaning? It affirms a simple relation between a potential function of the distances, and an integral of the relative velocities, in all cases where the forces are functions of the mutual distances of the parts, masses or atoms, but of these alone, so as not to depend either on the time or velocity, or any other element not of a local kind. This implies two main classes of possible laws, of which one, a very numerous class, depend on distance only, and the other class, far more numerous, are functions inclusively of other variables, and not of the distance alone. It is true for laws of the first class, not of the second. The equal possibility, in the abstract, of a great variety of laws of force is the basis and starting-point of the whole reasoning. The class included within the principle, as well as the class excluded, plainly admits of an immense variety of particular laws. The necessity of any one of these laws is thus strictly forbidden and excluded by the principle itself so highly praised.

But further, Gravitation is not the sole force in nature, and ponderable matter, as most students of physics allow, is not the only kind of material substance. The subject, it is true, has not passed wholly beyond the limits of debate. But the great majority of philosophers recognize an ether distinct from common matter, and those who affirm two or more varieties of it are more numerous than those who strive to account for light, heat, and electricity, by diffused ponderable matter alone. It may be viewed as almost, if not quite certain, that there is a self-repulsive ether, distinct from self-attractive matter, and yet somehow united with it in the closest way. How does this great fact of modern physics bear on the question now before us?

Now the bare existence of two kinds of material substance, distinct from each other, is a plain disproof of physical fatalism. There can be no reason, in abstract fate, why any atom should have the properties of matter, rather than of ether, or of ether rather than of matter. The atheistic doctrine, if there be matter alone, requires simply that each atom should have been its own creator and origin. But if matter and ether both exist, and have quite different properties, each of the trillions of atoms must have decided, at or before its own birth, which of the two kinds of being it would assume.

Again, if there be two kinds of physical substance, Matter and Ether, there must be three laws, at least, to determine their mutual action. To satisfy the known facts, these must all be distinct, and follow a different rate of variation. All reasons against the necessary nature of gravitation will thus apply, a second and a third time, to the other laws of cohesive force and of ethereal repulsion. But their threefold character adds a further element to the disproof of their inherent necessity. If one were necessary, it must exclude the two others. Necessity, like a Turkish sultan, admits of no rivals near the throne. Each atom would have to choose, not only whether it should be born, and whether it should be matter or ether, but which of these different laws of force it should for ever obey.

If we examine each of the dozen hypotheses, briefly named in the last chapter, the same reasoning will apply. The precise form of the argument may vary, but its essential elements are the same. No law or set of laws can be proposed, which offer the least sign of satisfying the great problem of physical change, except as a choice out of many alternatives, capable alike of being traced out into their consequences, but of which one alone proves to satisfy the actual phenomena. Observation and experiment would else be one immense paralogism, and all physical laws might be determined by the analyst or geometer in his study, without the need of any inductive process whatever. Those who hold such a doctrine ought to close our laboratories and observatories, and dispense with all the laborious methods of actual research, and set about solving all the mysteries of the universe by arithmetic and geometry alone.

The passage, then, by which Mr Spencer has replaced his earlier assertion, that gravitation is a necessary result of the laws of space, contains all the elements for a complete disproof of his whole theory. He now writes as follows:—

"If we cannot in thought see matter acting on matter through a vast interval of space which is absolutely void; on the other hand, that the gravitation...should be absolutely the same, whether the intervening space be filled or not, is incomprehensible. I lift from the ground, and continue to hold, a pound weight. Now into the vacancy between it and the ground is introduced a mass of matter of any kind whatever, in any state whatever, hot or cold, liquid or solid, transparent or opaque, light or dense, and the gravitation is entirely unaffected. The whole Earth, as well as each of the particles composing it, acts on the pound absolutely in the same way, whatever intervenes, or if nothing intervenes. Through eight thousand miles of the Earth's substance each molecule at the antipodes affects each molecule of the weight I hold, in utter indifference to the fulness or emptiness of the space between them. So that each portion of matter, in its dealings with remote portions, treats all intervening portions as though they did not exist, and yet recognizes their existence with scrupulous exactness in its direct dealings with them. We have to regard gravitation as a force, to which every thing in the universe is at once perfectly opaque in respect of itself, and perfectly transparent in respect of other things. While, then, it is impossible to form any idea of Force in itself, it is equally impossible to comprehend its mode of exercise." (F. P., p. 63, 3rd edition.)

Now it is quite true that the Law of Gravitation, when we meditate on it closely, does involve the facts here named, and others no less wonderful. Its very simplicity is apt to hide from us its real grandeur and mystery. There is nothing new or original in the passage which here replaces a very strange error. I have anticipated the remark in my "Treasures of Wisdom" thirty years ago. (Tr. W. pp. 50—52.)

What conclusion seems to follow from the truth of the Newtonian law, when once we grasp it in its full meaning? Every atom, in itself a mere nothing, a point or almost a point in space, needs to borrow what seems almost the Divine Omniscience, before it can fulfil the requirements of a rule so simple and so comprehensive. It is hardly possible to believe that any finite intelligence could solve this immense problem, in which the exact place, varying every moment, of trillions on trillions of atoms, decides the force exerted by each of them in turn. This problem of almost infinite complexity has to be solved anew, from hour to hour, and from moment to moment. Imagination faints under the load of a conception so vast and wonderful. The idea involves no contradiction. But it seems to compel the thoughtful mind, even in this lowest basement of physics, to look beyond second causes, and own the presence and working of One whose wisdom is infinite. The words of our Christian poet are almost forced upon our memory:

>   But how should matter execute a law,
>   Dull as it is, and satisfy a charge
>   So vast in its demands, unless impelled
>   To ceaseless action by some ceaseless Power,
>   And under pressure of some conscious cause?

The Law of Gravitation, when we reflect on it calmly, seems to force upon us one of two alternatives. The first is, that in all the motions which satisfy it material atoms are simply passive, and a Supreme mind, infinite in wisdom, and alone capable of solving the vast problem involved, is the sole real agent. The second is that the creation of matter implies a loan by the Creator, to points which are nothing in themselves, not only of a power of action and motion, but of some limited participation in his own cognizance of all localized being. For to obey the law, each atom should discern intuitively, with no outward medium of communication, its own distance, every moment, from every other material particle in the whole universe.

The original statement, that gravitation results by a fatal necessity from the laws of space, is withdrawn, and replaced by another, that the force of attraction is wholly unaltered by any amount of intervening matter. And hence it is inferred that we can form no idea of force, or the mode of its exercise. But the true inference is quite different. Gravitation, it implies, involves an immediate and not a mediate relation of all particles to each other. Each atom must have direct attraction or appetency to every other atom. As a fact, this is quite definite and comprehensible. But when we seek to resolve it into some other cause, there is none simpler, or even equally simple, of a physical kind, to which it can be referred. Whether we resort to ultramundane impact, or ethereal pressure, we have to devise a most complex machinery, composed of ethereal or ultramundane particles, exceeding in number almost infinitely those of matter, and needing fresh and more complex laws to account for their motions. They need to be so numerous as hardly to be distinct from a plenum, and so rare as not to interfere with each other, and move almost as in empty space. We seem, therefore, to reach in gravitation an ultimate law. If we try to replace it by some other physical explanation, we exchange the simple for the complex, the definite for the indefinite, the clearly conceivable for vague hypotheses; which, after all, fail to satisfy one main feature of the law, as deduced from all known experience, or the perfect transparency of all other atoms with regard to the action of each pair, however distant, taken alone.

The authority of Newton, in his letter to Bentley, has often been quoted to prove that gravitation, in his fixed judgment, was a mediate result of some other physical cause. He writes as follows:

"It is inconceivable that inanimate brute matter should, without the mediation of something else which is not material, operate upon and affect other matter without mutual contact, as it must do, if gravitation in the sense of Epicurus be essential and inherent in it....That gravity should be innate, inherent, and essential to matter, so that one body can act upon another at a distance, through a vacuum, without the mediation of anything else, by and through which their action and force may be conveyed from one to another, is to me so great an absurdity, that I believe no man, who has in philosophical matters a competent faculty of thinking, can ever fall into it."

The view, however, which Newton thus condemns, is not that gravity is physically immediate and ultimate, but that it is conceivable, alike in the absence of some physical medium, and of any spiritual and immaterial agent. He leaves the presence or absence of a physical medium, by which its cause might be carried one step further backward, an open question. But what he affirms is that, either such a medium exists, or else we need to recognise in it one of those "active principles, by which the things themselves are formed," and due to "the wisdom and skill of a powerful, everliving Agent; who, being in all places, is more able by his will to move the bodies within his boundless sensorium, than we are able by our will to move the parts of our own bodies." For God, he says, "has no need of such organs, He being every where present to the things themselves."

The law of gravity, if it be physically ultimate, shuts out and excludes the whole fatalistic theory. For who can believe that every atom, by some fatal inherent necessity, solves every moment a problem of all but infinite complexity, immensely beyond the powers of Descartes, Pascal, Newton, Euler, and every later analyst, and which needs for its solution, momently renewed, a perfect knowledge of the exact distance and position of every other atom, even to the farthest boundaries of the sidereal universe? Unless each solitary atom of matter, by some strange and fatal necessity, immensely outstrips and exceeds the highest scientific attainments of the greatest analysts and geometers, this Newtonian law must demand the ceaseless activity of a Being endowed with omniscient wisdom. Our thoughts can find no resting place or repose, till we look above all second causes to the sublime confession of the Psalmist: "Great is our Lord, and of great power; yea, and his wisdom is infinite "

In his Optics, query 21, Newton inclines to the view that gravity may perhaps be explained by the unequal pressure of an ethereal medium on opposite sides. But in query 28, a later passage, he seems to lean rather to an opposite view as his final conclusion; or that gravity is one of two or three laws or active principles, appointed by the Creator in the very act of creation, and capable of being resolved into nothing physically more simple, but only into His creative will. He writes as follows:

"A dense fluid can be of no use for explaining the phenomena of nature, the motions of the planets and comets being better explained without it.... And for rejecting such a medium, we have the authority of the oldest and most celebrated philosophers of Greece and Phœnicia, who made a vacuum and atoms, and the gravity of atoms, the first principles of their philosophy; tacitly attributing gravity to some other cause than dense matter. Later philosophers banish the consideration of such a cause out of Natural Philosophy, feigning hypotheses for explaining all things mechanically, and referring other causes to metaphysics. Whereas the main business of Natural Philosophy is to argue from phenomena without feigning hypotheses, and to deduce causes from effects, till we come to the very First Cause, which is certainly not mechanical. And not only to resolve the mechanism of the world, but chiefly to resolve these and such like questions. What is there in space almost empty of matter? And whence is it that the sun and planets gravitate to each other, without dense matter between them? Whence is it that nature doeth nothing in vain? And whence arises all that order and beauty which are in the world?...How come the bodies of animals to be contrived with so much art, and for what ends were their several parts? Was the eye contrived without skill in optics, and the ear without a knowledge of sounds? How do the motions of the body follow from the will, and whence are the instincts of animals?...And these things being rightly despatched, does it not appear from the phenomena that there is a Being incorporeal, living, intelligent, omnipresent, who in infinite space, as it were in his sensorium, sees the things intimately in themselves, and thoroughly perceives them; and comprehends them wholly by their immediate presence to himself? And though every true step in this philosophy brings us not immediately to the knowledge of the First Cause, yet it brings us nearer to it, and on that account it is to be highly valued.

"It seems to me, farther, that these particles have not only a 'vis inertiæ,' but also that they are moved by certain active principles, such as that of gravity, and that which causes the fermentation and the cohesion of bodies. These I consider, not as occult qualities, but as *general laws of nature, by which the things themselves are formed.* ...Now by the help of these principles all material things seem to have been composed of the hard, solid particles above mentioned, variously associated in the first creation by the counsel of an intelligent Agent."

One thing seems thus to have remained doubtful in Newton's mind,—whether gravity were an indirect result of a difference between opposite ethereal pressures, or an ultimate principle, one of two or three of the same kind, forming part of the definition of Matter in its first creation, and maintained by the direct agency of the omnipresent Creator. But of this he had no doubt, that all second causes, whether the steps be one or more, lead up to the great First Cause, and that this First Cause is not mechanical, but a true, living, intelligent, and omnipresent Mind; and that only by the presence and ceaseless activity of such a Being can the existence and permanence of gravity be really explained. The substitution of a blind Fate for the true and living God was, in his deliberate judgment, not less opposed to genuine science than to the deeper instincts of true piety and religious faith.

The first disproof of Physical Fatalism is found in the certain truth, that the Law of Gravitation, and other kindred laws of cohesive force, or ethereal repulsion, not yet precisely ascertained, can with no shew of reason be ascribed to a fatal necessity. The steps in the discovery of that law, which is already known, and the approaches made towards the detection of the rest, by the union of observation and experiment with mathematical reasonings, shew them plainly to be alternatives, out of a large variety of laws equally conceivable in themselves. Out of a multitude of possibles, these alone are proved by experience to be the real and actual laws of nature. Thus blind necessity can never account for them, but only the free choice of the great Author of the universe.

The same truth, however, is still more conspicuous, when we reflect on the variable elements of number, place, and relative position, without which not one of these various laws of nature would have real objects to which they could apply, and by means of which alone they can operate and exist.

What is implied, for instance, in the real existence of the Newtonian law? Every particle of matter attracts every other, or tends towards it, with a force proportional to the inverse square of the distance between them. Let us concede, for the sake of argument, that this law itself might depend on some kind of fatal necessity. But what of the distances? All these clearly are different now from what they were yesterday, and will be to-morrow. There is nothing whatever in the law to fix their amount, either now, or at any earlier time. At the most, it determines a connexion and relation between earlier and later positions, whenever those at some one point of time, past, present, or future, have been arbitrarily assumed.

The number of atoms of solid matter in a linear inch, from various phenomena, could scarcely be less than a thousand millions, and may be much greater. The radius of the sun is 400,000 miles, or 700 millions of yards, and 25 thousand millions of inches. Supposing the matter of the universe to amount only to eight cubes of which the side is the sun's diameter, or about sixteen times the sun's mass, the linear number of atoms would be $10^{11} \times 10^9 = 10^{20}$, and the cubic number greater than 10^{60}, or unity followed by sixty ciphers. This immense number, tripled, will give the coordinates, on which the relative places of all these atoms must depend. And the total number of forces, for all the pairs of atoms, will be $n\ (n-1)$ or unity followed by 120 ciphers, or in other words, a million multiplied into itself twenty times.

Again, it seems to result from all optical and electrical phenomena that a self-repulsive ether does exist, diffused through the whole universe as far as light extends. Its atoms also, in a linear inch, could hardly be fewer than a thousand millions. The radius of our solar system, half way to the nearest stars, is not less, probably, than a trillion of inches, and the range of distance accessible to our telescopes may be a thousand times greater still. From these data the number of ether monads would not be less than 4 followed by ninety ciphers, and their coordinates not less than 12 followed by the same number. And if all these monads act on each other, the number of forces, composing the total energy, would be much more than unity followed by 180 ciphers, or a million multiplied into itself thirty times.

Let us dwell only on the material atoms, of which the number must be more than a decillion, or a million multiplied into itself ten times. Let us suppose the three coordinates, x, y, z, disposed in a series, in the order of their values from any fixed origin. Since these differ by a finite interval, however small, each must admit of an infinite number of other values, still retaining its own place in the succession. So that infinity, first cubed, and then this cube multiplied into itself more than a decillion times, or to a power expressed by unity followed by sixty figures, will be an inferior limit for the number of possible variations of those coordinates, on which the amount and direction of the various accelerating forces must depend, even by this one simple law of universal gravitation.

What, then, is the amount of the fixity which the operation of the law implies, compared with the diversity which it cannot have the least tendency to explain or remove? On assuming any one set of original positions for all the atoms of the universe, the law will determine all their later motions and positions. Its effect is simply to hinder the lapse of time from increasing by one new variable the amount of variability which must already exist. The law does nothing at all to lessen the amount of indeterminateness involved in the positions at any one moment. It merely prevents this vast amount from being increased and multiplied by the countless number of successive moments of time. So that the amount of variability removed is to that which is still retained, and which no law of force can remove, in the ratio of unity to three times the number of atoms in the whole universe.

The doctrine of the Conservation of Energy, we have been lately assured, "binds nature fast in fate to an extent not before recognized." And if we make the very large and groundless assumption that no laws of action exist anywhere in the universe but the law of gravitation, and a few others of the same class, in which the force exerted by one unit on or towards another is a function of their distance alone, then the positions of all the atoms at any one moment would be linked indissolubly with their position at any other time, earlier or later, throughout all ages of the actual history of the world. That indeterminateness will be wholly excluded, which results from the lapse of time alone. But then the indeterminateness which still remains is infinite. The laws, even assuming their existence and unbroken validity, can do nothing to remove it, but the Almighty Lawgiver alone.

Why the number of atoms of matter, or monads of ether, should be exactly what it is, and neither less nor more; why they should occupy their actual positions, and not others partly or wholly different; why they should be moving with their actual velocities, and in their actual directions, when each of them plainly might have been in an infinite number of different places, and moving with an infinite diversity of other motions, are questions to which physical science can give no answer whatever. The notion that each atom fixed for itself, at some time or other, its own place, or that it was fixed for it by some other atom of a more commanding genius, is certainly a pseud-idea, and really unthinkable. In the weighty words of Newton, "blind necessity, which is certainly the same always and everywhere, could produce no variety of things." The contrast between parts of empty space, and solid, impenetrable atoms; or between points of mere space, and force-centres endued with powers of attraction and repulsion, or of appetency and self-motion; could spring from no fatal necessity, but from choice, will, and presiding intelligence alone. It compels us to rise above mere laws and formulæ to the conception of a First Cause, true, living, and intelligent, who by his powerful, all-creating word has called this wonderful universe into being. Whether worlds were first created in a comparatively finished form, as suns and systems, or a diffused nebulous ocean of matter and ether was endued with such laws of force, and its atoms placed in such positions, as that finished worlds would result from their action in some later age, the amount of choice involved and required will be just the same. The distance which separates such a scheme from fatal necessity is simply infinite and immeasurable. And the deeper we search into the nature of the laws which are found to prevail, the more light is thrown on the truth and grandeur of those simple sayings of the word of God: "He spake, and it was done; He commanded, and it stood fast." "By the word of the Lord were the heavens made, and all the host of them by the breath of his mouth." "They continue to this day according to thine ordinance, for all things serve Thee."

CHAPTER XI.

ON EVOLUTION.

The doctrine of Evolution is one main pillar, on which the philosophy of modern Materialism is supposed to rest. It has been said, in a sermon recently published, that this is the latest revelation which God has given to mankind; that it extends beyond the sphere of physics, and is opposed to the old conception of a world created out of nothing. The statements of the Bible, which contains earlier messages from the same divine source, are to be so interpreted and received as to agree with this later and fuller revelation. Whatever contradicts it must be laid aside, and held to be no part of the essence of Biblical teaching, but only of its separable husk, its historical and circumstantial form. For God cannot contradict himself. The doctrine of evolution is his latest voice, and men of science are the modern prophets, through whom it has been revealed to this age of mankind. We ought therefore to retain only so much of the teaching of the Bible as we can reconcile with this new and later revelation.

It must be very important, then, to see clearly what is this new doctrine, for which so lofty a position is claimed. For some, even among Christian Divines, have now come to look upon it as a test and touchstone, so certain, so firmly established, by evident proofs, that not only interpretations of Scripture, but direct statements of the Bible itself, are to stand or fall, to be received or wholly laid aside, as they agree or disagree with this great discovery, assumed to be one recent and firm result of scientific research.

What, then, is Evolution? It means properly an unwinding, unrolling, or opening out, of what existed previously in a more condensed or hidden form. Its simplest illustration is the process by which a thread of cotton or silk is unwound from the reel, until its whole length becomes clearly visible. The evolute of a curve is a second curve, formed by supposing a thread, wrapped on the first curve, to be gradually unwound. Thus in evolution there is no new existence. Something which exists already is seen in a clearer light, and becomes conspicuous to our eyes, so that one part of it is no longer hidden and concealed by another.

Out of this simplest meaning of the word a second has naturally arisen. Every real, actual being continues to exist through successive moments of time, and many fluctuations of continual change. These bring with them frequent variations of form, depending on the inward nature of each object, and the circumstances in which it is placed. The successive changes, speaking generally, cannot be foreseen. They are the subjects of observation and experiment. And hence these successive states of every permanently existing object are a kind of continual evolution. The cause is seen and known by its effects, and these are gradually disclosed. In this sense the whole course of history, physical and moral, may be viewed as one perpetual course and process of evolution.

So far there can be nothing new and original in the

doctrine. It is simply one mode of expression for the teaching of reason and common sense in every age. If permanent beings, with definite powers and faculties, do exist, and those powers cannot be known intuitively, but in their consequences, then all nature and all providence must be a constant unfolding of those secret powers. They will evolve themselves from day to day, and from year to year, through successive generations.

Evolution, then, implies and assumes the existence of something which evolves and unfolds itself, or is unfolded by another. It can be no substitute for creation, but only its consequence. What excludes creation is not evolution, but either non-existence or self-existence. If all atoms of matter existed of themselves from all eternity, with powers of attraction and repulsion, motion and change, there would then be no room for a material creation, and all the positions these atoms may assume would be a kind of evolution. Again, if plants and animals have no individual being, and are merely a certain or uncertain number of material atoms in some special arrangement, there need be, and can be, no creation of what does not exist. On this hypothesis nothing exists but the component atoms, and we deceive ourselves with names, when we suppose that each plant and animal has any proper being, for which an act of creation would be required.

Again, if we suppose that in generation there is no new individual, which then first begins to be, but merely the continuance of one or both parents, or that the parents themselves have no proper being of their own, the whole course of animal and vegetable changes will be resolved into a kind of evolution. It may be either the evolution of some specific life, diffusing itself into many moulds, or simply of matter, assuming new and complicated forms.

But if we retain the usual and natural view, that individual men and animals do really exist, then the term evolution will apply to the successive forms or phases the same individual assumes; but creation or generation, not evolution, will be the proper term for the origin of these individuals themselves.

The Scripture doctrine on this subject seems to be precise and clear. The first plants and animals of each kind owe their origin to a direct act of Divine power. This is what we call creation. But later plants and animals owe their origin to a power bestowed on the first parents, to bring forth after their kind, or produce an offspring like themselves. This is procreation or generation, and is distinct alike from simple, immediate creation, and from evolution, or the unfolding of the properties of some being which already exists.

The Evolution of modern philosophers, when used to replace and exclude the Scripture doctrine of Creation, is thus made up of two dogmas. First, in the field of mere physics, that all the atoms of lifeless matter are uncreated, self-existent, and eternal. And next, in physiology, that generation is no commencement of a new individual being, but only the varied arrangement of an uncertain number of pre-existing material monads; or else a new distribution of some vital force or power, of which the same total amount, but differently arranged or divided, existed long before.

In Mr Spencer's First Principles seven chapters are occupied with an exposition of the nature and supposed laws of Evolution. The sameness of the principle, as applied to the changing forms of lifeless matter, and the appearance and birth of plants, animals, and men, is not proved at all, but assumed as self-evident. Thus two distinct questions of high importance are confused and blended into one. In this chapter I shall examine the doctrine in reference to lifeless matter alone, and in two others, that follow, I shall extend the inquiry to living things and persons.

Evolution, with reference to material objects, may be used either in a literal or metaphorical sense. In the former, it implies the existence of something composed of many parts, wound up or closely compacted together; and means some process by which they are unwound, or separated without a total severance, and thus brought out more clearly into view. In the metaphorical sense it denotes the successive effects, in course of time, of powers, causes, or objects endued with such powers, and the phenomena to which they give birth. Thus a drop of water may be said to evolve itself, when it passes into steam and evaporates, or a spark of fire when it kindles a destructive conflagration.

The term Evolution, in Mr Spencer's First Principles of Philosophy, receives a different and almost opposite meaning. It is defined to be "a change from incoherent homogeneity to coherent heterogeneity, accompanying the dissipation of motion, and the integration of matter." Such a definition evidently needs a paraphrase, which will be found in the following statements of the same work.

The premises of the whole reasoning are the three principles already examined, that matter is indestructible, motion continuous, and force persistent. These are taken as the fundamental data of the problem. The problem itself is "the law of the continuous redistribution of matter and motion." "The question to be answered is, what dynamic principle, true of the metamorphosis as a whole, and in its details, expresses the ever changing relation."

Philosophy has to formulate the passage of things from the imperceptible to the perceptible, and back into the imperceptible again. "The formula must be one comprehending the opposite processes of concentration and diffusion. The change from a diffused imperceptible state to a concentrated perceptible state is an integration of matter and concomitant dissipation of motion; and the change from a concentrated perceptible to a diffused imperceptible state is an absorption of motion, and concomitant disintegration of matter. These are truisms. Constituent parts cannot aggregate, without losing some of their relative motion; and they cannot separate, without more relative motion being given to them. We are not concerned here with any motion, which the components of a mass have with respect to other masses, but only with the motion they have with respect to one another. The axiom we have to recognize is that a progressive consolidation involves a decrease of internal motion, and that increase of internal motion involves a progressive unconsolidation.

"All things are growing or decaying, accumulating matter or wearing away, integrating or disintegrating. Since there is no such thing as an absolutely constant temperature, every aggregate is every moment progressing towards either greater concentration or greater diffusion."

"The processes thus everywhere in antagonism, and everywhere gaining now a temporary, now a more or less permanent triumph, we call Evolution and Dissolution. Evolution under its simplest aspect is the integration of matter and the dissipation of motion; whilst Dissolution is the absorption of motion, and the concomitant disintegration of matter."

The main element, then, in Evolution, according to Mr Spencer's scheme of philosophy, is concentration, or the transition of matter from a more diffuse to a more condensed and perceptible form. The second, hardly less important, is the change from a homogeneous to a diversified and heterogeneous structure. He traces this change, successively, in the sidereal systems, the geological stages of the earth's history, the growth of plants and animals from the seed or embryo, the comparison of extinct and living species, and the physical and social features of the various tribes and races of mankind. It is further traced in the history of language, and the fine arts, and the greater variety of occupations in civilized societies. And the general conclusion is drawn that "along with a passage from the incoherent to the coherent there goes on a passage from the uniform to the multiform. Such at least is the fact, wherever evolution is compound....The entire mass is integrating and simultaneously differentiating from other masses, and each member of it is also integrating and differentiating from other members."

Applied to the whole system of the universe, the theory takes this general form. Evolution is the progress from a vague, incoherent, and nearly homogeneous nebula, of vast and almost infinite extent, to finished sidereal systems, suns, planets, satellites and comets, air, sea and land, and countless varieties of living things, of which every one is composed of many unlike parts, and a complex organization. Dissolution is the reverse process, by which all these worlds dash themselves together, shatter themselves into pieces at the last, and thus resolve themselves once more into diffused nebulosity, mist, and darkness, while the wearisome oscillation of change goes on for ever.

Here an objection will at once arise. Evolution properly denotes unfolding or expansion. How can it then be taken, without great violence, to denote the opposite process of condensation? Mr Spencer notes the difficulty, and replies to it as follows:

"Evolution has other meanings, some of which are incongruous with, and some even directly opposed to, the meaning here given to it. The evolution of a gas is literally an absorption of motion and disintegration of matter, which is exactly the reverse of what we here call Evolution. As ordinarily understood, to evolve is to unfold, to open and to expand, to throw out or emit, whereas, as we understand it, the act of evolving implies that its matter has passed from a more diffuse to a more concentrated state, or has contracted. The antithetical word, Involution, would much more truly express the nature of the process, and would indeed better describe its secondary characters. We are obliged, however, notwithstanding the liabilities to confusion that must result from these contradictory meanings, to use Evolution as antithetical to Dissolution. The word is now so widely recognized as signifying, not indeed the general process above described, but sundry of the most conspicuous varieties of it, and certain of its most remarkable accompaniments, that we cannot now substitute another word. All we can do is, carefully to define the interpretation to be given to it."

Evolution, in strictness of speech, is clearly an unwinding, unrolling, or expansion, of something which exists already in a more compressed, contracted, or less developed form. If it has been used, in some cases, to denote what is directly opposite, it seems a very strange remedy for the partial abuse thus introduced into scientific nomenclature, to make it universal, to turn the meaning of the phrase upside down, and define it as the process of concentration and material aggregation. Philosophers often overrate greatly their own power over language. When they choose to employ words in an artificial and non-natural sense, they can seldom escape the danger of confusing their own thoughts, but never avoid that of perplexing, misleading, and mystifying their readers. Who could suppose, when they hear this doctrine of evolution proclaimed as a great scientific discovery of our times, by which the old Christian doctrine of Creation is dispensed with and set aside, that the real doctrine intended is not one of evolution at all, but of involution, or of some natural tendency in the whole universe, and in all its parts, to contract and condense, to become less evolved or diffused, and to increase in seeming complexity? This is just the same as to affirm that a skein of thread is unwound, when we fold it up into the least compass, and entangle its parts together in every possible way.

But let us waive this verbal objection, however real and important, and examine the doctrine itself. How far does it embody a true and exact principle of dynamical reasoning, or practical observation? Evolution, strangely used to denote concentration, consists in the integration of matter and dissipation of motion. Dissolution, the reverse process, that is, expansion, consists in the absorption of motion and concomitant disintegration of matter. These are said to be truisms. But when we look at them closely, we shall find them not to be truths at all, but certain and demonstrable falsehoods. Thus the whole structure of supposed discovery, based on this double definition, must fall to the ground.

The first premise of the theory is in this statement: matter when extremely rare and diffused, contains the greatest amount of motion, and it parts with this motion, in proportion as it condenses, and assumes some permanent structure.

Now this axiom, if words are taken in their only strict and proper sense, is wholly untrue. It exactly reverses the real truth. Our actual universe, according to the nebular theory, has undergone a process of condensation for immense ages, and it certainly now contains a vast amount and variety of motion. But the primitive nebula, if there be any soundness in the theory, must have been once in a state of nearly absolute rest. The only logical ground on which the hypothesis can rest is the greater simplicity of ascribing all motion to original forces, compared with the conception that one part only of the motion is due to those known or unknown forces, but that the rest was imparted directly and supernaturally. The only real presumption for such a starting-point, instead of the creation of finished systems, when we contemplate lifeless matter alone, consists in the greater simplicity of the view that forces alone were directly created, and that all actual motion is the result of those forces; compared with the notion that forces were imparted in creation, but that some motions were also communicated directly from the Creator's hand. The laws of mechanics point to the conclusion, that the regular orbits and rotations of the planets, and the nearly globular shapes of the larger celestial bodies, would result in the course of ages from the law of universal attraction, modified by cohesive forces which certainly exist, but of which the precise laws are unknown, if these acted for long ages on a universe of matter widely diffused, and originally at rest. (*Scripture Doctrine of Creation*, pp. 98—100.)

Thus the ideal starting point in the Nebular Theory is

not a state of maximum motion, but its exact opposite, one of perfect rest. And it is just as easy to conceive such a state of rest when matter is very widely diffused as when it is gathered into dense masses. We know, in fact, that the component parts of all the bodies that surround us are in a state of incessant and rapid motion, while the first state of the primitive nebula may have been, and if the doctrine be well founded, must have been, the presence of countless motion-generating forces, but a total or nearly total absence of actual motion.

How can we explain, in this exposition of the evolution theory, so direct and complete an inversion of the real truth? It grows out of the dangerous practice of using terms in a non-natural sense. What has been done before with the word evolution, is now repeated with the word motion. Energy is first used ambiguously, to express either force or motion in turns, or to speak more precisely, their integrals. The integral of force is styled Potential, that of motion, Kinetic energy. Next, motion is made to replace energy in either sense, and is styled either actual or potential. "An aggregate that is widely diffused," Mr Spencer remarks, "contains a large quantity of motion, actual or potential, or both." But this potential motion is not motion at all, but a pure fiction, or the possibility that something now at rest may hereafter come to move. Potential Energy, again, is an integral of possible forces that will come into play in the event of certain changes of the mutual distance. And thus, as a caravan in a pathless desert may come by gradual deviations to return on its own track, so it is in the basis of this theory of evolution. It is mistaken for an axiom that a primitive nebula will have a maximum of motion, when it would have no motion at all, but absolute rest; and that the present universe having passed through long stages of evolution, must contain much less motion than the nebula in which it had its birth, when the actual amount of motion is almost infinite, and the primitive amount, in an ideal nebular system, none whatever.

The second part of the fundamental axiom is not less erroneous. The great change in Evolution is said to be "an integration of matter, and a concomitant dissipation of motion." Here motion is looked upon as a distinct and independent substance. The matter or substance of a system may all condense, and be aggregated together, while the motion leaves it, like an aura or effluvium, or delicate odour, and exhales away into empty space. But this is a pseud-idea, unthinkable and impossible. Motion is not a thing or substance, but an accident, quality or relation, of something that changes its place, or moves. If matter, then, moves and passes away, this must be a process of diffusion and expansion, and not of concentration. On the other hand, if matter aggregates and the motion diminishes also, this is simply an approach nearer and nearer to a state of perfect rest.

The two elements, then, which together make up Mr Spencer's definition of evolution, do not agree. They contradict and exclude each other. The case supposed, in which the universe is formed out of diffused nebulous mist, is one in which attractive forces must prevail. Condensation would then ensue, and increase for long ages. The collective motion, however, would not diminish, but grow and increase. Now heat is atomic motion. It is one of the simplest and best known laws of physics, that heat ensues upon condensation, while expansion of every kind is attended with the lowering of temperature, or the production of cold. Thus the doctrine laid down in the above passages is made up of two great errors. Motion cannot possibly be dissipated unless some part of the matter or substance is dissipated also, for plainly it cannot subsist alone. And again, concentration of matter, in every case which has occurred or is likely to occur, must be attended, not by a diminished amount of motion, but by its large and very abundant increase.

Evolution, in fact, as defined by Mr Spencer in the field of physics, is neither more nor less than a very misleading and inappropriate name for the process of cooling. A mass condenses when it cools. At the same time, one part of the motion produced by condensation passes away, through the medium of the ambient ether, and is transferred to other parts of the system. This could not be without the intervention of some other substance than the sensible mass, which ought to be taken into account as one part of the whole system. The heating or cooling of any piece of matter is really a change affecting a compound system both of matter and ether. But since matter is self-attractive, and ether self-repulsive, in the composite system a double process of central condensation, and superficial expansion, may easily go on side by side. Thus motion may seem to leave the central portion, and be dissipated, when we fix our thoughts on the condensing mass of ponderable matter, and overlook the more subtle and less perceptible elements of ether and etherealized matter, which form an essential part in the whole process of change. The poetry of Milton is thus a nearer approach to the truth than Mr Spencer's definition of evolution:

> Swift to their several stations hasted then
> The various elements, earth, flood, air, and fire;
> And this ethereal quintessence of heaven
> Flew upward, spirited with various forms.

The mistake in the definition may be presented in another light, by comparing it with the other main basis of the physical theory of nature, the Conservation of Energy. This may be summed up in the general statement that Energy is of two kinds, Potential or the Energy of Position, and Kinetic or the Energy of Motion; and that under the laws of force, actually known, or reasonably conjectured, the sum of the two is constant, and one loses as much as the other gains. That the doctrine may be true, Energy of Position, with attractive forces, must be reckoned greatest when the distances are greatest, and with repulsive forces when the distances are smallest. When repulsive forces predominate, the motion must thus increase with the amount of dispersion and diffusion. But when attraction predominates, the motion will be greatest when the system is condensed, and the energy of position is the least.

The definition of evolution, adopted by Mr Spencer, exactly reverses this doctrine. It makes the Potential Energy, or Energy of Position, and the Kinetic Energy, or collective motion, both increase together, and become greatest when the system is most diffused; and both diminish together, when it contracts and condenses into a smaller space. And thus the integration or condensation of matter, which by the law of Conservation of Energy must be connected with the increase of collective motion, is strangely affirmed to be the concomitant of its dissipation and utter loss.

Such is the confusion which readily ensues, when, instead of definite laws of force, disclosed by comparing exact and careful experiments with mathematical reasonings, we try to solve hard questions with a picklock, and build up a scheme of physics by a plentiful use of abstract metaphysical phrases alone. We may soon lose ourselves, as has plainly been done in the present instance, in an intellectual jungle, overgrown with pseud-ideas and contradictions, where no genuine outlet or clear pathway can be found.

It is useless and impossible to analyse in detail the two hundred pages, in which this fundamental mistake is developed, and applied to all moral and social, as well as purely physical changes, under the common title of evolution. I will examine one section only, near the opening of the whole exposition, and which reads as follows (pp. 282—284):—

"Thus far we have supposed one or other of the two processes to go on alone, or an aggregate to be either losing motion and integrating, or gaining motion and disintegrating. But though every change furthers one or other of these processes, it is not true that either process is ever wholly unqualified by the other. For each aggregate is at all times both gaining motion and losing motion. Every mass, from a grain of sand to a planet, radiates heat to other masses, and absorbs heat radiated by other masses, and in so far as it does one, it becomes integrated, and in so far as it does the other, it becomes disintegrated. Ordinarily in inorganic objects this double process works but unobtrusive effects. Only in a few cases, among which that of a cloud is the most familiar, does the conflict produce marked and rapid transformations....If drifting over cold mountain tops, it radiates to them much more heat than it receives, the loss of molecular motion is followed by increased integration of the vapour, ending in the aggregation of it into liquid, and the fall of rain. Here, as elsewhere, the integration or disintegration is a differential result."

"In living aggregates, especially those classed as animals, these conflicting processes go on with great activity, under several forms. There is not merely the passive integration of matter, which results from simple molecular attractions, but there is an active integration of it under the form of food. Animals produce in themselves active internal disintegration by absorbing external agents into their substance. While, like inorganic aggregates, they passively give off and receive motion, they are also active absorbers of motion latent in food, and active expenders of that motion. But it remains true that there is always a differential progress towards either integration or disintegration. During the earlier part of the cycle of changes integration predominates, and there goes on what we call growth. The middle part is usually characterized, not by equilibrium between the two processes, but by alternate excesses of them. And the cycle closes with a period in which the disintegration, beginning to predominate, puts a stop to integration, and undoes what integration had originally done. At no moment are assimilation and waste so balanced, that no increase or decrease of mass is going on. The chances are infinity to one against these opposite changes balancing each other; and if they do not, the aggregate as a whole is integrating or disintegrating."

"Everywhere, and to the last, the change going on at any moment forms a part of one or other of the two processes. While a general history of any aggregate is definable as a change from a diffused, imperceptible state to a concentrated perceptible state, and again to a diffused imperceptible state, every detail of the history is definable as part either of one change or the other. This, then, must be that universal law of redistribution of matter and motion, which serves at once to unify the various groups of changes, as well as the entire course of each group. The processes thus everywhere in antagonism, and gaining now a temporary, now a more permanent triumph, one over the other, we call Evolution and Dissolution."

Let us examine this statement, first, as it refers to lifeless matter. It includes two assertions, almost self-evident. Every material aggregate, as a general rule, either gains or loses, is subject to partial waste from its surface, or increases by deposition of some fresh matter. The coasts of islands are worn away, in some parts, by the waves of the sea, and in others gain by the silting up of sand. Also every mass, in general, is either growing hotter or cooler, and the temperature very rarely, if ever, remains quite the same. So far the statements are true. But they plainly carry us a very little way towards a philosophical theory and explanation of all the changes in the universe.

But two other statements follow. "Every aggregate is at the same time both gaining and losing motion." This is not only untrue but impossible. The motion of an aggregate must be the sum total of the motion of its parts, and either external or internal. Its external motion is that of its centre of gravity, and this cannot at the same moment be moving both faster and slower than before. It may do either, but cannot do both at once. Again, the internal motion of the aggregate is the sum total of all the relative motions of its separate parts or atoms. If n be the number of its parts, the number of these relative motions is $n\,(n-1)$, or their square diminished by the number itself. Each of these elements separately may either increase or diminish. But their sum total, which is the internal motion of the aggregate, cannot do both at once. It can only do either in turn. That there should be both gain and loss in the motion of the aggregate, that is, the aggregate motion, at the same moment, is strictly impossible.

The other principle affirmed is that any aggregate grows cooler when it is increased, and becomes hotter when it is diminished. But this has no ground either in experience or dynamical theory. Increase of mass by new matter, and increase of density the mass being unaltered, are confounded under the common term, integration. But in both alike the usual effect is exactly the reverse of what is affirmed in this attempted theory of evolution. It is not the loss or dissipation of motion, but its increase. The mass tends to become hotter, not cooler, by the change. So true is this, that one theory of solar heat, popular only a few years ago, sought to explain it by the continual impact of meteors falling on its surface, and thus arrested in their previous orbits. By this change the mass and the temperature would plainly be increased together.

The theory, when applied to living creatures, or as Mr Spencer styles them, "the living aggregates classed as animals," leads to further paradoxes. They are said to produce in themselves active internal disintegration, by absorbing external agents into their substance. Eating food is thus represented as a process, not of evolution, but of dissolution, and not an increase but a diminution of the corporeal mass. Again, they are said to be "active absorbers of motion, latent in food, and active expenders of that motion." If, however, each animal is merely an aggregate of the chemical atoms that compose its bodily structure, and nothing more, how can it act at all, either to absorb food, or to expend what has been thus absorbed by voluntary acts of motion? The separate atoms may act on each other, and on neighbouring atoms, by laws of gravitation and chemical affinity, even when life has ceased. But vital acts, or any difference between life and death, on such a hypothesis, are wholly unaccountable. The word, starve, is used in our language ambiguously, to denote either deprivation of food, or of the heat essential to comfort and life. The strange result of Mr Spencer's doctrine, applied to animals, is to make these really two opposite extremes, that exclude each other. Eating food undoes the effect of waste, increases the total mass or weight, and should thus be a process of integration. Its proper concomitant, by the theory, must be dissipation of motion, in other words, loss of heat and starvation by cold. On the other hand, the wasting of the body by lack of food lessens the aggregate. And this, by the same theory, implies absorption of motion, that is, the increase of the vital heat. To avoid starvation in one sense of the word, the proper course would thus be to practise it in the other, and abstinence from food would be the scientific prescription for the safe performance of an arctic voyage.

In fine, the definition of Evolution, as a physical process, which Mr Spencer has laid down, and endeavoured to confirm through two hundred pages of detailed exposition, exactly reverses the real connection of the two elements of which it is composed. Aggregation may be of two kinds, an increased closeness or density of a material mass, unchanged in amount, or increase of the mass itself by arrested impact, or accretion on the surface. But in both cases the direct and almost necessary concomitant is increase of the collective motion, and not its loss or diminution. On this very account, because the motion is increased, and the temperature raised, there will often ensue, as the second state, distinct from the first, and following after it, a process of cooling, or the radiation or conduction of heat from the mass which has been heated to surrounding bodies. This secondary and indirect result of the aggregation is confounded in the theory with that which is primary and direct. It is further overlooked that this secondary result cannot occur without a process the direct opposite of the first. It requires, not the concentration, but the dissipation and expansion of some part of the constituents of the whole compound system. Motion cannot radiate into space alone, apart from something that moves. And thus the central condensation, by which motion is increased, is followed by a further change of dissipation at the surface, either of matter extremely rare, surcharged with ether, or of the ether alone.

But even when this great error has been removed, and replaced by the opposite truth, a still deeper and broader objection remains, which lies against the whole theory, when made the basis for a philosophical creed, and explanation of all the changes of the universe. Its professed aim is to supply "the law of the continuous redistribution of matter and motion," and by its help to explain, not physical changes alone, but the higher problems of vegetable, animal, and rational life, and all complex and mysterious relations of mental thought or of human society. All is summed up in the one word, Evolution, of which the antithesis is Dissolution. But this Evolution, when we examine it closely, is neither more nor less than the process of condensation, which results from the acting of the law of universal attraction, and other similar laws of force, on a vast and immeasurable number of material atoms, very widely dispersed originally through the depths of space. Now if the laws themselves were known, the results to which they would lead are definite, and it would be the province of dynamics to unfold them. It is true that their relation is just the opposite of what Mr Spencer affirms, and links condensation inseparably, not with the dissipation and loss, but the increase of motion. But when this mistake has been removed, what remains of the doctrine? Simply the fact that masses of matter, under the influence of attractive and cohesive forces, tend to become denser at the centre, and more diffuse at the circumference or periphery, and at the same time to develop a great amount of motion, most of which, soon or late, will be likely to take the form of swift rotation or revolution. There may here, perhaps, be a partial key to the mode in which suns, planets and satellites, have received their actual forms, through the laws and conditions of place assigned by the Creator in the beginning of his great work. But this process of condensation does not give us the least help towards a true conception of life, or the birth, growth, and structure of the countless varieties of the animated or even the vegetable world. Indeed it seems demonstrable that no such results could ensue from laws of mechanical force alone. These atoms, wonderful exceedingly within their own appointed limit of being, beyond that limit are utterly powerless. They can move, but they cannot think or feel. The laws they fulfil without deviating need little short of omniscience to satisfy them for a single moment. Each atom must either be able to divine, each instant, the place and distance of every other atom in the universe, to effect an almost infinite summation of these various tendencies to be obeyed, and that without a moment's cessation or pause, or else be guided passively by the hand and secret wisdom of the Almighty Creator. They can pull or push, and change their relative positions, with amazing docility, or still more amazing intelligence. But here their power ceases, so far as human science can extend its view. The results are new places, altered velocities, and varied forces tending to fresh change, but nothing more. No trace of life, choice, or thought, of pleasure or pain, hope or fear, right or wrong, moral good or evil, can be found in such a universe. For these require higher and nobler gifts to be received from the great Source of all created being. If these are withheld, the whole universe would contain nothing more than dense, whirling balls of lifeless matter, or scattered and floating patches of nebulous vapour and confusion, and remain a dreary and barren wilderness, a waste and desolation for evermore.

CHAPTER XII.

ON HETEROGENEITY.

THE doctrine of Evolution, in 'First Principles,' is more completely expressed in the following formula, which sums up the statements and reasonings of eight successive chapters of the work. "Evolution is an integration of matter, and concomitant dissipation of motion; during which the matter passes from an indefinite, incoherent homogeneity to a definite, coherent heterogeneity; and during which the retained motion undergoes a parallel transformation."

Simple evolution, according to the theory now to be further examined, consists mainly in the aggregation or condensation of masses or material systems, and is said to have for its invariable concomitant the dissipation of motion. Compound Evolution introduces the further idea of secondary redistributions. As the first character is called integration, the second is styled differentiation. These terms are not used in their mathematical sense, in which one is the exact converse and reversal of the other. The two processes are held to be constant companions, and go forward side by side. The first leads to greater density, the second to varieties of shape, structure, or quality, in the condensing mass. It is illustrated from all departments of science. The following paragraph will be a sufficient specimen of the nature of the theory.

"A growing variety of structure throughout our sidereal system is implied by the contrasts that indicate an aggregative process throughout it. We have nebulæ that are diffused and irregular, and others that are spiral, annular, spherical. We have groups of stars, the members of which are scattered, and groups concentrated in all degrees, down to closely packed globular clusters. We have groups differing in number, from those containing several thousand stars to those containing but two. Among individual stars there are great contrasts, real as well as apparent, of size, and from their unlike colours, as well as unlike spectra, numerous contrasts in their physical state are to be inferred. Beyond which heterogeneities in detail there are general heterogeneities. Nebulæ are abundant in some regions of the heavens, while in others there are only stars. Here the celestial spaces seem almost void of objects; and there we see dense aggregations, nebular and stellar together.

"The matter of our Solar System, during its aggregation, has become more multiform. The aggregating gaseous spheroid, dissipating its motion, acquiring more marked unlikenesses of density and temperature between interior and exterior, and leaving behind from time to time annular portions of its mass, underwent differentiations that increased in number and degree, until there was evolved the existing organized group of sun, planets and satellites. The heterogeneity of this is variously displayed. There is the immense contrast between the sun and the planets, in bulk and in weight, as well as the subordinate contrasts of like kind between one planet and another, and between the planets and their satellites.

There is the further contrast between the sun and the planets in respect of temperature; and there is reason to suppose that the planets differ from one another in their proper heats, as well as in the heat which they receive from the sun. Bearing in mind that they differ also in the inclinations of their orbits, the inclinations of their axes, in their specific gravities, and their physical constitutions, we see how decided is the complexity wrought in the Solar System by those secondary redistributions which have accompanied the primary redistribution."

The same fact, of a growing diversity, is next traced in Geology, and the successive real or supposed stages of the cooling of the earth. "Between our existing earth, and the molten globe out of which it was evolved, the contrast in heterogeneity is sufficiently striking." It is traced, further, in the changes of vegetable and animal growth. As to the comparison of fossil and existing species, the evidence is owned to be doubtful, and suffices neither for proof nor disproof. "Yet some of its most conspicuous facts" it is said "support the belief that the more heterogeneous organisms, and groups of organisms, have been evolved from the more homogeneous." The principle is then applied to human history. The truth that mankind, as a whole, have become more heterogeneous is so obvious as scarcely to need illustration. Every work on Ethnology, by its divisions and subdivisions of races, bears witness to it. In passing from its individual to its social forms, the same law is still further exemplified. Society in its lowest forms is a homogeneous aggregation of individuals of like powers and like functions. Every man is warrior, hunter, fisherman, toolmaker, every woman performs the same drudgeries. Yet very early we find an incipient differentiation of the governing and the governed. At the same time there arises a co-ordinate species of government, that of Religion. "For many ages religious law continues to contain more or less of civil regulation, and civil law to possess more or less of religious sanction; and even among the most advanced nations these two controlling agencies are by no means completely differentiated from each other." "The ever increasing heterogeneity in the governmental appliances of each nation has been accompanied by an ever increasing heterogeneity in the governmental appliances of different nations; all of which are more or less unlike in their political systems and legislation, in their creeds and religious institutions, in their customs and ceremonial usages."

The same principle of growing diversity is further traced in human language, and the various arts of life, Painting, Sculpture, Architecture, Poetry, Music and Dancing. Everywhere "the entire mass is integrating, and simultaneously differentiating from other masses; and each member of it is also integrating, and simultaneously differentiating from other members."

But while Evolution is a change from the homogeneous to the heterogeneous, it is also a transition from the indefinite to the definite. This is traced out in reference to the structure of the body in health, compared with the results of disease, and to healthy political developments, in contrast to the abnormal and confusing effects of all merely revolutionary change. Again, "change from indistinct characters to distinct ones was repeated in the evolution of planets and satellites, and may in them be traced much further." "With progressive settlement of the space-relations, the force-relations become more settled. The exact calculations of physical astronomy shew us how definite these force-relations now are; while their original indefiniteness is implied in the extreme difficulty, if not impossibility, of subjecting the nebular hypothesis to mathematical treatment."

But Evolution is said to imply further, not only an integration of matter, and a dissipation of motion, and a differentiation of the matter, by which it becomes more heterogeneous, but also a differentiation of the motion, which is not dissipated and lost, but still retained. This aspect of the doctrine is thus explained:

"If concrete matter arises by the aggregation of diffused matter, then concrete motion arises by the aggregation of diffused motion. That which comes into existence as the movement of masses implies the cessation of an equivalent molecular movement. While we must leave in the shape of hypothesis the belief that the celestial motions have thus originated, we may see, as a matter of fact, that this is the genesis of all sensible motions on the earth's surface. The molecular motion of the ethereal medium is transformed into the motion of gases, thence into the motion of fluids, and thence into that of solids, stages in each of which a certain amount of molecular motion is lost, and an equivalent motion of masses is gained. It is the same with organic movements. Certain rays proceeding from the sun enable the plant to reduce special elements existing in gaseous combination around it to a solid form, enable it to grow, and carry on its functional changes. And since growth, equally with the circulation of the sap, is a mode of sensible motion, while the rays expended in generating it consist of insensible motions, we have here too a transformation of the kind alleged. Animals carry this a step further....Both the structural and functional motions which organic Evolution displays are motions of aggregates generated by the arrested motion of units....In all cases where the incident forces do not vary as the masses, every new order of aggregation initiates a new order of rhythm. Witness the conclusion drawn from the recent researches into radiant heat and light, that the molecules of different gases have different rates of undulation. In proportion as any part of an evolving whole segregates and consolidates, and in so doing lessens the relative mobility of its components, its aggregate motion must obviously acquire distinctness. A finished conception of Evolution we thus find to be one which includes the redistribution of the retained motion, as well as of the component matter" (pp. 382—395).

The number of facts of various kinds, in physics, physiology, human and social life, brought together in these six chapters of the First Principles (Bk. II. ch. 13—18) is at first very impressive. We seem surely to have caught the glimpse of something like a law of nature of wide extent, and almost universal. But when we examine the statements more closely, the illusion will disappear. Instead of a physical or social law we find only a phrase, and clasp a shadow.

How can a tendency of different aggregations to become unlike each other be, in any proper sense, a law of nature, or any key to the course of physical change? For all the masses compared are supposed to be changing together. Let A and B be two such masses, nearly resembling each other. They tend, it is said, to grow unlike. What can be more indefinite than such a rule or guide of action? Does A seek to become unlike what B now is, and B unlike what A now is? Then both variations might be of the same kind, and the new forms resemble each other as much as the old ones. Or does each aim to become different from what the other is about to become? Such a law of mutual repulsion from shapes or states not yet in existence, is inconceivable and incredible. That the course of physical changes may actually issue in growing diversity is quite conceivable. But this can only be as the consequence of some definite laws of force or change, and itself can be no real law, or any key to the true definition of those changes in which this diversity of result appears. All genuine laws of nature are both definite in themselves, and lead to definite consequences. A tendency of like things to become unlike is a principle the most vague and indefinite conceivable, and can never rank as a proper law of nature.

Again, Evolution is said to be a progress from the indefinite to the definite. But on the theory of Physical Fatalism, every thing must be equally definite from first to last. A leading writer of this school has said that the British fauna and flora of this day was potentially present in the primitive nebula, and might have been certainly predicted from it by a sufficient intelligence. Such a view is clearly absurd, if there were any indefiniteness of the original or any later stage of evolution, or indeed in any part of the process of change from first to last.

The true course of physical progress, if we accept the nebular theory, is not what Mr Spencer affirms, from prevailing likeness to unlikeness, from the homogeneous to the heterogeneous, and from the indefinite to the definite. It is from likenesses and unlikenesses, not perceptible by the senses, but by the reason only, to those which are perceptible alike by the senses and the reason. It is from a definiteness so complicated as to be imperceptible and incalculable, to one which may be partly recognized by touch and sight, and which admits also of partial calculation. Aggregates of matter may be seen and handled. Atoms or monads can neither be seen, felt, nor heard, and are objects, not of our senses, but of the understanding alone. There is thus no real progress from the homogeneous to the heterogeneous. Likenesses and unlikenesses coexist in the primitive nebula, as they coexist in masses inorganic and organic, and in worlds and systems. It cannot be shewn, even, that the relative proportion of unlikenesses to likenesses is increased by this development. But the first are invisible and impalpable, and must be apprehended by the pure reason. The last come within the range of our senses. And if we dwell in thought on the sensible unlikenesses alone, since there are none of these in a system of uncondensed atoms, it is easy to fall under the illusion that progress from likeness to unlikeness is a kind of universal law of physical development.

Let us consider, first, a system of matter not aggregated or integrated, a primitive nebula. In what sense is this homogeneous? We may consider each of three hypotheses, that of small solid atoms of various shapes, sizes, and forms, or of spherical atoms, or of force-centres. The first is the view of Leucippus, Democritus, Epicurus, Lucretius, Newton, and most atomists; the second that of Mosotti, Challis, and several modern physicists; the third is that of Boscovich, Exley, Bayma, and many others. On the first view there is a great amount of likeness in atoms of the same kind. They are manufactured articles of the same type, unalterable by combination, and differing in place alone. But, on the other hand, the unlikeness of atoms of different kind is fixed and unalterable. If we assume only ten kinds of primitive atoms, the number of unlike pairs that can be formed is ten times greater than that of the pairs of the same kind, and the number of triplets in which all differ in kind will be 72 times greater than of triplets in which all agree. If the primitive varieties were a hundred, then the unlike pairs would be a hundred times, the unlike triplets almost ten thousand times, more numerous than the pairs or triplets of like atoms.

On the first hypothesis the primitive atoms differ both in form and size, on the second in size only. But this plainly admits, in the abstract, of infinite varieties. If we assume all to be of the same size, for the present object the hypothesis becomes identical with that of force-centres.

Here all the atoms are of two or three kinds only, those of matter and ether, or in some varieties of the doctrine, of two ethers. If there be two kinds of substance only, the like pairs will exceed the unlike, unless the number of each is equal. But if n be the number of atoms of either kind, which are selected for later aggregation, the same will be the number of varieties, of which two will consist of atoms of the same kind, and the rest of different kinds and in different proportions. Thus, the unlike varieties surpass the like varieties in proportion to the number of atoms chosen, whether they are still dispersed or closely aggregated together.

Besides this, in the constitution of such a primitive nebula, there are two fundamental contrasts or unlikenesses, which are inseparable from the conception of it, as much as the resemblances of the atoms or force-centres. The first, assuming it to be finite, is the contrast between the whole finite space which contains the system, and the unoccupied, empty space, which lies beyond. And if we venture to endue the material atoms with hyperphysical powers, and ascribe to them the potency of every form of life, this fundamental contrast and unlikeness would be rendered still more complete. The other contrast is between each monad or force-centre, and the space which lies immediately around it. If we take a geometrical spherical surface of any definite size, and conceive it moved from place to place through the nebulous and diffused mass, with every fresh position the set of atoms included will be different, and differently related to the including sphere, so that the unlikenesses of these diverse systems will be more numerous by far than the total number of atoms in the whole nebulous system.

All these primitive unlikenesses, though present in the system from the first, and clearly cognizable to our reason, are imperceptible to the senses. It is not surprising, then, that philosophers should fall into the illusion of conceiving that unlikeness is generated by the course of development, when the real fact is that it existed equally before. The real change consists in likenesses and unlikenesses, which once were latent and insensible, being brought within the range of sensible observation.

This is the second part of the answer, which disproves Mr Spencer's elaborate theory, that Evolution is a tendency from the homogeneous to the heterogeneous. For in the process of aggregation the sensible likenesses are multiplied and increase, no less than the sensible diversities. Thus, in astronomy, every sun has a marked resemblance to every other sun, planet to planet, and satellite to satellite. The orbits of all the planets of our system, and their successive orbital revolutions and axial rotations, strictly resemble each other. As a general rule in these developments, the likeness is primary, and the unlikenesses are secondary. Every name, in fact, is the condensed expression of the resemblance which exists in a whole class of things to which that name applies. So that the growth and enlargement of language is a direct proof that likenesses are multiplied and increased with every step in the actual process of material change.

CHAPTER XIII.

ON FORCE AND LIFE.

ONE main feature in the modern doctrine of Evolution is the attempt to resolve life and thought into special varieties of the organization of matter, and physical change. Solar force, under certain conditions, is held to transform itself into mental phenomena, and these in turn are transformed into voluntary movements. There is a manifest relation, we are told, between the amount of emotion and the muscular action induced, from the erect carriage and elastic step of exhilaration up to the dancings of immense delight; and from the fidgetiness of impatience to the almost convulsive movement accompanying great mental agony. The inference is drawn that we are indebted to the solar radiations, as the grand reservoir, for all the subtle and complex manifestations of force which are evolved in human thought and society. Physiology, humanity, politics, morals, are to be viewed as only special modifications and developments of physical science.

It is a hard task to explain, without a Creator, the existence, laws, and changes, of a vast physical universe of lifeless matter. To explain in the same way, by uncaused, uncreated laws of evolution the immense variety of vegetable, animal, and human life, must be far more difficult still. Yet this is the task which many modern theorists in biology propose to themselves. It may well be of deep interest to examine the process and order of their attempts at its solution.

In entering on a problem of such extreme difficulty, the first essential must plainly be to have a clear conception of the nature of Life, the main subject of the investigation. No genuine theory can be reared on a pedestal of nescience and confusion. Clear definitions must precede, if careful and exact reasonings or observations are to follow. How far do modern speculations on Life satisfy this first condition of all genuine science?

Mr Spencer's "Principles of Biology" are the formal application of the general principles of the philosophy of Evolution to the subject of Life. They present this aspect of the theory of evolution in its best known, and possibly its ripest form. Several other definitions of Life are proposed, and rejected as defective or inadequate. Two others, as more exact and complete, are given in their stead, and the second of these is preferred as a working definition. It is "the continual adjustment of internal to external relations." It is "the definite combination of heterogeneous changes, both simultaneous and successive, in correspondence with external coexistences and sequences."

Here a first question will naturally arise. Is this new definition simpler, clearer, and easier to understand, than the terms for which it is to be a substitute, life, vital force, and vitality? These last are rejected, almost with scorn, by many recent writers of the school of materialism. Is the proposed substitute a change for the better? Does it bring into clearer relief the characters by which life is differenced from lifeless matter? Has it the slightest resemblance to that law of gravitation, the great example of genuine physical discovery, in which the terms themselves are strictly defined, and can be made the basis of a whole course of definite reasoning? Does it not rather illustrate the saying of the witty French author, that one chief use of words was to conceal thoughts? Great elasticity is an excellence in the steel of a balance-spring, but is by no means a quality to be admired in a philosophical definition.

Let us endeavour to analyze this definition, and resolve it into the various elements of which it is composed.

First, Life is a combination of changes. It is not the cause or source of changes, but those changes themselves. How many of these changes, then, are needed to satisfy the definition? Through how many changes must a living man or animal have passed, in order to be really alive? And again, of what are they to be the changes? Of some millions or billions of atoms, which have existed for countless ages, and have been changing through every moment of their existence?

But Life is a combination of these changes? How is this at all possible? How can these changes combine at all, since one state of this set of atoms must have ceased before the next comes into being? Can past, present, and future changes combine together, and all exist at the same moment? This is surely a thing impossible.

Again, if Life is a combination of various changes, who or what is to combine them? The theory excludes any reference to a Creator. The term, persistence of force, is introduced to avoid the risk of suggesting an idea, foreign and strange to this Lucretian philosophy, that of a Divine Preserver and Sustainer of all things. Not the living plant or animal. The definition recognizes no such existence, but seems purposely framed to exclude it. Do these changes, then, combine themselves? Do successive changes all exist before they combine, or combine before they exist? Either alternative is unthinkable. Or is the combination nothing more than the simple fact of their successive occurrence? But what claim can such a series have to the title of a combination? The definition brings before us a number of successive phenomena, making up a total, called life, but nothing at all which can combine them, or tie them together.

A third element follows in this definition. Life is a definite combination of changes. But by whom or what is this combination defined? What is there to sever these changes from millions on millions of others, adjacent to them in place, and coexisting with them in time, which it is meant to exclude? How can a series of vital changes possibly be defined, unless we first recognize the existence of some unit, some living individual thing or power or person, to which they all belong?

These changes, again, are said to correspond with external coexistences and sequences. In the other and more brief definition, external and internal relations are named in contrast to each other. Here a fourth difficulty must arise. What can be the meaning of these terms, internal and external? If a living thing is neither more nor less than a number of atoms in a certain arrangement, at given distances, and having certain given velocities, whence can these conceptions of outness and inwardness arise? These epithets, internal and external, introduce by stealth and in secret that idea of a living unit, with a defined limit to the range of its powers, which the theory refuses openly to recognize, because it would be fatal to the whole course and tenor of its reasoning.

These changes, which are to constitute Life, when they have been combined without any combiner, and defined in the total absence of any Power competent to define them, are also said to be simultaneous and successive. But these can be no special characteristics of vital changes. They are common alike to all things, living and lifeless. A million of atoms cannot fail to have simultaneous changes, since they all coexist, or successive changes, since they all exist in time. But then they are also heterogeneous, or unlike in kind to each other. Instead of throwing any light on the subject, this added epithet renders it still darker than before. For vital changes must surely have this common feature, by which they resemble each other, and differ from other phenomena, that they are the changes in some living creature or thing. A living organism, it is true, will usually be more complex than an equal bulk of lifeless matter. Yet surely the changes of the sun, as revealed by the spectroscope, and those of the earth, apart from all life, in its seas, coasts, rivers, atmosphere, clouds, tides, and currents, are heterogeneous in a very high degree. Their variety, complication, and diversity, surpass the limits of human imagination.

A second main idea is next introduced, to supply the void which must be felt in the first part of the definition, when taken alone. These changes, in the combination of which Life is to consist, must be "in correspondence with external coexistences and sequences." But here our perplexities are only increased. How can changes correspond with sequences, that is, with phenomena which do not coexist with them, but come into being only when those changes are at an end? How can there be external coexistences when no living organism has yet been defined, or severed from the whole coexistent universe? What is meant by correspondence? All coexisting changes, of whatever kind, vital or non-vital, must correspond in time. The changes of a falling stone correspond with the forces of all the elements of the earth, which pull it downward, and with momentary changes of the air, through which it passes, and which it disturbs in its fall. The changes of every planet in its orbit correspond with the successive actions of the central force of the sun, with successive perturbing forces of the other planets, and with ethereal vibrations, by which it is made visible, whenever there are eyes to observe its course and position.

All these terms, combination, definite, external, correspondence, coexistence, imply what is not expressed openly, because irreconcileable with the purely material or mechanical notion of life, although essential to any true definition; that is, the presence of a living something, exercising vital forces, that lead to vital actions, distinct in kind from the mechanical and chemical movements of lifeless matter. A leading Positivist has proposed this third definition, that "Life is a series of definite and successive changes of structure and composition, which take place within an individual without destroying its identity." So far as this makes Life consist in the vital changes themselves, it lies open to the same decisive objections as the one already examined. But the ideas of individuality and identity are here plainly recognized, and bring it nearer to the truth than his own more complex definition, for which Mr Spencer sets it aside. Let its clauses be only inverted, and it will perhaps form a very near approach to a just definition. "Life is the identity of one and the same individual being, endued with powers by which it originates a series of definite and successive changes of structure and composition, in some organization with which it is intimately combined."

Let us now endeavour to approach to some definition of Life, at once in harmony with the facts of science, and with the natural and instinctive feelings of mankind. Three things seem, then, to be implied as its coessential elements: an individual that lives, in contrast with the many atoms of the living organic structure; a vital force, which attracts and appropriates, and builds up into an organism suitable materials, rejecting what is unsuitable; and a plan, aim, or structure, proper to each different kind of life, from which this double process of vital attraction and repulsion derives its special characters, and whereby it is differenced from the atomic forces of lifeless matter. This provisional definition, if still somewhat obscure or incomplete, will at least enable us to see more clearly the great defects of that which Mr Spencer has proposed, and in which every one of the three essential elements is either contradicted or ignored.

Life, then, is that force or power of some living individual existence, whether man, animal, plant or germ, by which it can attract into union suitable material, and repel or reject the unsuitable, in agreement with some plan of living structure, or external life-work, peculiar to each specific form and type of life.

First of all, life implies and requires some living thing or creature, distinct from the limbs of an organic structure, or its circulating fluids, or the adhesive molecules of a gelatinous mass. Common sense affirms this plainly with regard to the highest forms of life, and even far down in the zoological scale. A man, a lion, and even a bee, a butterfly, or a worm, is not a series of changes, combined in some inexplicable way, occurring in a hundred limbs or a million million molecules, and corresponding with other changes in "the environment," or the other countless molecules around them. A man is a living, rational intelligence. A lion is a living, sentient agent. Each bee, insect, butterfly, worm, is one and not many, however great the number of atoms of which its body is composed.

Schelling has defined life as "a tendency to individualism." This is indeed to put a plain and simple truth in a philosophical masquerade. Individuality is not an asymptote, and life a curve which approaches closer and closer to it. Life implies that individuality is already present. An insect or plant or animal is an individual something that lives.

The objections which have been raised to the essential individuality of life are drawn entirely from its lowest and most obscure forms, or microscopic animalcules, invisible to the naked eye. These are exceedingly minute in size, their numbers immense, their modes of reproduction very simple, and the laws of that reproduction very difficult to trace. But it surely reverses the first lesson of genuine philosophy to have recourse to the most obscure corner of a wide and important science for its definitions. This is to interpret day by night, and clear sunshine by mist and darkness. In all the higher forms of life, with which mankind have been familiar for long ages, the marks of individuality are clear, decisive, and irresistible. The principle, established here by so large and wide an induction, ought to be our guide when we descend into that lower region, where, from the minuteness of the microscopic forms, and their answering multiplicity, the distinctive characters of life seem almost to disappear wholly from view. The notion of "widely extending sheets of living protoplasm" is one of those guesses in the dark, which are sometimes, for a brief period, mistaken for profound philosophical thought. A mere conjecture, contradicted and disproved by later observers, is used to set aside a conclusion which results, almost irresistibly, from millions on millions of the most familiar and patent facts in the whole range of human experience.

Again, Life implies some special force or power, in that which lives, to attract suitable materials into some related structure or organism, and to repel or reject what is unsuitable. Thus it will agree with mechanical or chemical action in the general fact of the presence of a certain attractive and repulsive power. But it differs from these both negatively and positively. Negatively, it does not extend, like the force of gravity, to immense distances, but is confined to matter at small, and almost insensible distances from the living structure. And positively, it does not attract or repel by a reference to mere distance alone. It includes a kind of selection or choice, with reference to some prescribed and preexistent plan or type of structure. The degree or amount of this choice seems very limited in the lowest class of living things, but grows fuller and larger, as we rise higher in the zoological scale.

The remarks of Dr Beale, the able microscopist, on the features of life even in these lowest forms, are very striking and important. He writes as follows:—

"One characteristic of every kind of living matter is spontaneous movement. This, unlike any movement of non-living matter yet discovered, occurs in all directions, and seems to depend on changes in the matter itself, rather than on impulses from without. I have been able to watch the movements of small amœbæ under a magnifying power of 5000 diameters. Several of these were less than a hundred thousandth part of an inch in diameter, and yet were in a state of most active movement. The alterations in form were very rapid. They might be described as minute portions of very transparent material, exhibiting the most active movements in various directions, in every part, and capable of absorbing nutrient materials. A portion of what was one moment at the lowest point would pass in an instant to the highest. One part seemed, as it were, to pass through the other parts, while the whole mass moved, now in one, now in another direction. What movements in lifeless matter can be compared with these?"

The idea of spontaneity seems to result from a survey of the lowest, most minute, and purely microscopic forms of life, as well as to reveal itself directly in human consciousness, and to be indicated in all the various kinds of animal life most familiarly known. The idea that the action of a cat in seizing on a mouse, or a hawk on a pigeon, is merely solar light and heat transformed, and answers simply in kind to the *vis viva* of a bullet fired from a gun or pistol, is so strange and unnatural, that it is hard to conceive how it can be interwoven into a scheme of philosophy, or find credence with any reasonable mind.

A third characteristic of life and living action is the presence of some specific type, either in the structure of the plant or animal, or in the products and direction of its activity, on which all the attractive and repellent influences of vital energy depend. This action may be classed generally under three heads; growth and nutrition, to develop the embryo into a perfect individual; external structure, as in the nests of birds, the reefs of coral, or the cells of the beehive; and reproduction, by which the species is continued and multiplied. These functions differ immensely in different species of animal life, while there are common analogies which run through the whole series. The attempt to get rid of the idea of purpose or design, so as to refer all the acts of living creatures to mechanical laws and processes alone, is revolting to the common sense of mankind. Under learned phrases, it strives vainly to conceal an unusual degree of self-contradiction and logical absurdity. The hive bee aims at constructing its cell, and then at storing it with honey. The bird uses much skill in building its nest, and in preparing it for the future process of incubation. The mason does not seem to follow more truly the design of the architect, than insects of various kinds satisfy the outlines of some plan or type, in forming their habitations, which has been appointed to them from the beginning.

These three characteristics of Life, individuality of being, active vital power, and the selective nature of this power, with reference to some determinate type of internal structure, or external products of instinctive labour, seem to be incorporated in the very conception itself. Even theories which aim to set them aside, and would resolve it into nothing more than an evolution of mechanical and chemical force, are compelled, in spite of themselves, to introduce them surreptitiously at every turn. Thus in Mr Spencer's Principles of Biology we meet with the following statements, side by side with the definition in which all these conceptions are denied and disappear.

"The growth of an animal depends on the abundance and the size of the masses of nutriment, which its powers enable it to appropriate" (p. 113). "Much depends on how many individuals are competing for the food." Birds "cannot support themselves without relatively great efforts" (p. 114). "Men and domestic animals, overworked while growing, are prevented from attaining ordinary dimensions." "A complex animal, capable of adjusting its conduct to a greater number of contingencies, will be better able to evade damage" (p. 119). "Animal organisms have a certain power of selective absorption, which adapts the proportions to the need of the system" (p. 122). "The crocodile undergoes considerable exertion only during brief contests with prey" (p. 126). "Selective assimilation illustrates a general truth." "A cell differs from the rest, and initiates the developmental changes" (p. 153). "The highest animals repair themselves to a very small extent, mammals and birds only in the healing of wounds. The power of reproducing lost parts is greatest where the organization is lowest" (p. 173). "By selective assimilation the repair of organs is effected." "Groups of compound units have a power of moulding fit materials into units of their own form" (p. 177). In a lizard, "the organism, as a whole, exercises such power over a newly forming limb, as makes it a repetition of its predecessor." The living particles, composing one of the fragments of a begonia leaf, "have an innate tendency to arrange themselves into the shape of the organism to which they belong" (p. 180). "We may use the term, organic polarity, to signify the proximate cause of the ability which organisms display, of reproducing lost parts."

In all these passages the admission seems clearly to be made, that each animal is a living individual; that it possesses, as a whole, certain powers of action; that in embryos some microscopic part originates the development of the rest; that this power or force includes a faculty of selection; and that the aim or tendency of this elective polarity, is to reproduce, as in the case of a lost part, or even, in some vegetables, of a severed leaf, and in cases of waste and repair, to sustain and perfect, that specific type to which the individual belongs.

This term, organic polarity, has been repeatedly used by advocates of material evolution as the sole key to all natural changes, to form a substitute for the ostracised idea of living power. A comparison is drawn between the growth of crystals, and that of plants, animals, and men. So one writer affirms that "in a voltaic battery we have the nearest approach man has made to an experimental organism." This analogy, then, requires a few words of notice. In reality the phenomena of crystalline structure, when a probable hypothesis of atoms and their laws of force has been made, may be easily explained by attractive and repulsive forces alone. If chemical atoms are the first step in the order of composition from centres of force, or simple monads, they will of necessity be polar. This polarity, also, must be strongest, as a general rule, in the case of low atomic numbers. Every apex, in the ultimate chemical atom, will determine lines of greatest permanent cohesion. The structure of the whole crystal will thus be determined to one, out of a small and definite number of crystalline types or forms. The directive power will result from a combination of three elements, the number of the monads which make up the compound or chemical atom, their polyhedral position, and the geometrical conditions of space. A square, for instance, must have four corners, and other squares, approaching it, will experience differing attractions in its own plane, or in a plane at right angles, and in the direction of its four corners or its four sides. If the nearness is such that the force becomes repulsive, the differences will still be the same. But plainly a selective power, by which a lizard repairs a lost limb, or a begonia leaf, when planted, becomes a new begonia, is of quite a different nature. For it implies the preexistence of a definite type of organic being, and a power, impulse, or effort to attain it, by expansion and assimilation, under fixed conditions.

Life, if these remarks are just, involves three elements; the existence of living individuals, active vital power in each of those individuals, and a specific type or form, towards which this vital force tends continually. In lifeless matter we have attraction and repulsion, depending on the distance, and exercised towards all matter without discrimination. In vital action, we have both attractive and repulsive forces, but with discrimination, so as not to depend on distance alone. Let us now consider the bearing of this new element on that theory of force, which forms the main basis of the fatalistic theory of evolution.

The doctrine of a fixed, invariable quantum of force, the basis of the theory, is not true, even when we deal with physical forces alone. For Potential Energy, the main element of the problem, is not force in existence or activity at any moment, but an integral of all the forces that would be exercised between each pair of atoms, in passing from their actual distance either to infinity, to coalescence, or to a neutral distance, in the three cases of force purely repulsive, purely attractive, or changing from attraction to repulsion. In the second case the doctrine must wholly fail, since there can be no definite quantum in the sum of an almost infinite number of quantities, each infinite. And even in the third case, which seems to be the actual one, the elements of the potential, being in number nearly equal to the square of the number of atoms in the universe, depend for their realization on as many conditions, which exclude each other, and of which not one pair, out of millions of millions, can possibly be fulfilled together. It is a sum total formed out of the forces that would be exerted, or the velocities generated, by each pair of atoms if they existed alone, and passed by their attractive force from their actual to the neutral distance. Let n be the number of atoms in the whole sidereal system. Then $\frac{1}{2} n (n-1)$ is the number of hypotheses required, that the elements of the total potential may be separately realized; and all of these hypotheses are mutually exclusive, and no two of them can be fulfilled together.

The doctrine, then, of an invariable total of existing force, or even of potential force, or potential force increased by the amount of motion, is baseless and deceptive, even when the forces are of the kind which the theory of conservation of energy requires. As a help to calculation, and a useful dynamical theorem, a sum of hypothetical possibilities, and not of present realities, it applies to all cases in which the forces are functions of the mutual distances only, and to these cases alone. If there are any forces which are functions of the velocity or of the time, the system ceases to be dynamically conservative, and the theorem fails.

How, then, is the doctrine affected by the presence of Life, and of living or vital forces? Does each living unit exercise on its own organism, and on matter around it, a force depending solely on the distance? The very opposite is plainly true. Not only is the action on any part of the body widely different from that on a foreign unconnected body at the same distance, but an element of selection and discriminating choice makes its appearance, which cannot be reduced to a formula of distance, and of remoteness or nearness alone. And thus in a complex system, into which both mechanical and vital forces enter, it is highly probable, if not absolutely certain, that the theorem of the conservation of energy, even when confined to its true limits, and not perverted into a great cluster of metaphysical contradictions and falsehoods, will no longer apply.

All the reasonings, in the First Principles, on the question whether life can be included under a formula of mechanical force, have one fatal and evident defect. They prove, with a superfluous profusion of examples, that all exercises of vital activity are connected with mechanical changes, and depend on certain physical conditions. But they do not prove that the results depend on those physical conditions alone, and that no element of choice, ideal type, or spontaneity, enters into the results that follow. That a certain function varies when x varies, is no proof that its values depend on the variations of x alone. It may be a function of two or three, or many variables, and its values in that case may differ widely, when either variable in turn is constant, and does not vary.

Let us take one passage for a specimen of this faulty reasoning. "Plant life is all directly or indirectly dependent on the heat and light of the sun....Each plant owes the carbon and hydrogen of which it is mainly composed to the carbonic acid and water in the surrounding air and earth. The carbonic acid and water must, however, be decomposed before their carbon and hydrogen can be assimilated. To overcome the powerful affinities which hold their elements together requires the expenditure of force, and this force is supplied by the sun. In what manner the decomposition is effected we do not know. But we know that when, under fit conditions, plants are exposed to the sun's rays, they give off oxygen, and accumulate carbon and hydrogen.... Thus the irresistible inference is, that the forces by which plants abstract the materials of their tissues from inorganic compounds, the forces by which they grow and carry on their functions, are forces that previously existed as solar radiations.... While the decomposition effected by the plant is at the expense of certain forces emanating from the sun, which are employed in overcoming the affinities of the carbon and hydrogen for the oxygen united with them, the recomposition effected by the animal is at the profit of those forces which are liberated during the composition of those elements.... There is a tolerably apparent connexion between the quantity of energy which each species of animal expends, and the quantity of force which the nutriment it absorbs gives out during oxidation.... The transformation of the unorganized contents of an egg into the organized chick is altogether a question of heat. Withhold heat, and the process does not commence. Supply heat, and it goes on while the temperature is maintained, but ceases when the egg is allowed to cool." (F. P., pp. 208—211.)

Facts of the kind adduced in this passage, and which might easily be multiplied a hundred fold, are a convincing proof that light and heat, derived from the sun, are one condition in the development of vegetable and animal life. But they furnish no proof whatever that vital action is only solar force transformed, and that the results are due to mechanical heat alone. Their true evidence is exactly opposite. When two causes conspire in producing a given result, and it is unattainable without their concurrence, it is a gratuitous sophism to ascribe it to one only, and to consider the other a mere accidental concomitant. The plant cannot appropriate the carbon and hydrogen, which build up its substance, nor separate them from the oxygen, except by the help of solar heat and sunlight. But neither will the light and heat of the sun, without the vital action of the plant, effect this decomposition. The egg cannot be hatched without a certain amount of warmth or heat. But no amount of solar light and heat will turn a lump of chalk into a bird of any kind, however long it may shine upon it. The vital force can work only under definite physical conditions. But this can never prove that it is simply a product and result of those conditions. Life, in all animals, can act upon things without, only through the body of the animal, its organic structure. Now the mechanical and chemical laws which belong to each animal body, even when life has ceased, do not the less belong to it while the animal lives. A corpse and the living man will have an equal weight in a pair of scales, and the chemical atoms and elements that belong to each will be the same. So that Life, in all its various forms of activity, must ever be mixed up with physical conditions; and the other laws of nature, which belong to material physics, are not suspended, because a different and higher law mingles with them, and modifies their working, so as to produce effects wholly beyond the reach of mechanical agencies alone.

CHAPTER XIV.

ON NATURAL SELECTION.

THE modern theory of Physical Fatalism or Evolution undertakes to explain the origin of all living and extinct species of plants and animals, and of the countless individuals that belong to them, without any need to recognise the presence of intelligent design, or the hand of an all-wise Creator. There is thus involved in it an entire reversal of one of the deepest instincts of human thought, common to the learned and the unlearned of every age.

The chief support of this theory is found in a supposed discovery of a few recent authors, called the law of Natural Selection. To this results the most prodigious have been ascribed. It is an Atlas, which has to bear on its shoulders the main burden of the origin and construction of the whole world of animated being.

The term itself is plainly a misnomer. For the chief aim of the theory, of which it is held to be the main support, is to dispense with and exclude the idea of intelligent design in the whole range of the material creation. A much more appropriate name is "the survival of the fittest," suggested by Mr Spencer in his Principles of Biology. One or two extracts from Mr Darwin's work on the Origin of Species will serve to explain its meaning.

"The preservation of favourable variations, and the rejection of injurious variations, I call Natural Selection. No complex instinct can be produced by it, except by slow and gradual accumulation of numerous slight but profitable variations....The mind cannot possibly grasp the full meaning of a hundred million of years. It cannot add up or perceive the full effect of many slight variations, during an almost infinite series of generations....By this theory all living species have been connected with the parent species of each genus by differences not greater than we perceive between varieties of the same species in the present day. And these parent species, generally extinct, have in their turn been similarly connected with more ancient species. And so on backward, always converging to the common ancestor of each great class, so that the number of intermediate and transitional links between all living and extinct species must have been incomparably great."

"Natural Selection, on the principle of qualities being inherited at corresponding ages, can modify the eggs, seed, or young, as easily as the adult....We may account for the distinctness of whole classes from each other, as birds from all other vertebrate animals, by the belief that many unusual forms of life have been utterly lost, through which their early progenitors were formerly connected with those of the other vertebrate classes."

"There is a power, Natural Selection, always intently watching each slight accidental alteration in the transparent layers of the eye, and carefully selecting each alteration, which may in any way produce a distinctive image....On this principle we must admit that all the organic beings which have ever lived upon the earth may have descended from one primordial form. If Natural Selection be a true principle, it will banish the belief of a continued creation of new organic beings, or any great and sudden modification of their structure."

"I see no limit to the amount of change, to the beauty and infinite variety of the adaptations, which may be effected in course of time by Nature's power of selection.... I can see no reason to doubt that Natural Selection might be most effectual in giving the proper colour to each kind of grouse, and keeping that colour, once acquired, true and constant. We must believe that these tints are of service to these birds, and to insects, in preserving them from danger. We do not see the transition through which the wings of birds have passed. But what special difficulty is there in believing that it might profit the modified descendants of the penguin, first to flap along the surface of the sea, and ultimately to rise from its surface, and glide through the air?...A well developed tail having been formed in an aquatic animal, it might come to be subsequently worked in for all sorts of purposes, as a fly-flapper, a prehensile organ, or an aid in turning....I see no difficulty in a race of bears being rendered, by Natural Selection, more and more aquatic in their structure, with larger and larger mouths, till a creature was produced as monstrous as a whale....I can see no reason to doubt that female birds, by selecting, during thousands of generations, the most melodious and beautiful males, according to their standard of beauty, might produce a marked effect....Individual males have had, in successive generations, some slight advantage in their weapons, means of defence, or charms, and have transmitted these to their offspring....I see no difficulty in Natural Selection preserving and accumulating variations of instinct to any extent that was profitable. It is thus, I believe, that all the most complex and wonderful instincts have originated....A slight modification of structure or instinct, correlated with the sterile condition of certain members of the community, has been advantageous to the community. Consequently the fertile males and females of the community flourished, and transmitted to their fertile offspring a tendency to produce sterile members having the same modification. I believe this process has been repeated, until that prodigious amount of difference which we see in several insects between fertile and sterile females of the same species, has been reached....By the long continued selection of the fertile parents, which have produced most neuters with the profitable modification, all the neuters come to have the desired character."

The first and most natural impression, I think, when we read such statements, is that they are an experiment how far the passive credulity of some readers will extend. But the doctrine has gained such eminent proselytes, and been so widely received, that we are bound to examine its claims with gravity and due respect. Sir C. Lyell, reversing his earlier convictions, has espoused it warmly in the following words:

"To many this doctrine of Natural Selection seems so simple, when once clearly stated, and so consonant with known facts, that they may have difficulty in conceiving how it can constitute a great step in the progress of science. Such is often the case with important discoveries. But to assure ourselves that the doctrine is by no means obvious, we have only to refer back to the writings of skilful naturalists, who theorised on the subject in the earlier part of the century, before the invention of this new method of explaining how certain forms are supplanted by new ones, and in what manner these last are selected out of innumerable varieties, and rendered permanent."

In his Principles of Biology, Mr Spencer goes further

than Sir C. Lyell in his praise of the theory. The work on the Origin of Species, he says, has proved it to the satisfaction of nearly all naturalists. Its truth, when once enunciated, is so obvious as scarcely to need proof. To Mr Darwin "we owe the discovery that Natural Selection is capable of producing fitness between organisms and their circumstances. He has the merit of appreciating the immensely important consequences that follow. He has worked up an enormous mass of evidence into an elaborate demonstration that this preservation of favoured races in the struggle for life is an ever-acting cause of divergence among organic forms. He has traced out the results involved in the process with marvellous subtilty, and shown how hosts of facts otherwise inexplicable are explained by it. He has proved that the cause alleged is a true cause, and one which we continually see in action."

The claims advanced on its behalf rise still higher. In the presence of this grand discovery, not only the Christian doctrine of special acts of creation, but even the Nomotheism, which would substitute for these a continuous operation of creative power, a tendency to development, at first supernaturally bestowed, must be consigned to the cemetery of exploded theories. He speaks of it as follows:

"In whatever way formulated, or by whatever language obscured, this ascription of organic evolution to some aptitude naturally possessed by organisms, or miraculously imposed upon them, is unphilosophical. It is one of those explanations which explains nothing, a shaping of ignorance into the semblance of knowledge. The cause assigned is not a true cause. It is unrepresentable in thought, one of those illegitimate symbolic conceptions, which cannot by any mental process be elaborated into a true conception. The assumption of persistent formative power, inherent in organisms, and making them unfold into higher forms, is no more tenable than that of special creations, differing only by the fusion of separate unknown processes into a continuous unknown process."

Natural Selection is thus, by its admirers, placed on the same level with the Newtonian law of gravitation. It is a true cause, in contrast to a mere hypothesis, a proved result of induction from all the known facts, and not a mere guess, requiring the assumption of the past existence of unknown and unproved phenomena exceeding a thousandfold in amount all the facts that are really known. Let us try to carry out the comparison thus suggested a little further. Its true analogy, I believe, will be found in the vortices of Descartes, and not at all in the Newtonian law of gravitation.

First, the phrase, Natural Selection, is thus explained and defined by its author. "Every one knows what is meant by such metaphorical expressions, and they are almost needful for brevity. It is difficult to avoid personifying Nature. But I mean by Nature the aggregate action and product of many laws, and by laws the sequence of events, as ascertained by us."

If this definition be correct, the principle can have no possible claim to be a grand philosophical discovery. It can include only a copious registration of the observed " sequence of events " that may form the raw materials of some theory hereafter to arise. The essence of the doctrine, however, does not consist in a registration of "ascertained sequences," but in the free invention of conjectural antecedences, through millions of years or ages, before man was born, or experiments and observations could be made.

The theory, it is plain, involves several postulates of a

very definite kind, apart from which it can have no meaning whatever. The first of these is the existence of Life, and a great variety and multitude of living things, as a fact known by experience. Natural Selection does nothing to explain this, and leaves it as complete a mystery as the common view of creation. The attempt of Mr Spencer to throw light on it, by a new definition of life, only leaves the darkness as profound as before. The second postulate is the fact of reproduction, or that plants and animals produce other plants and animals like themselves. This also is a result of experience on the widest scale, and its simplest and most comprehensive expression is found in the words of the creative fiat in Genesis, "whose seed is in itself after its kind." This fact also the new theory has to assume, but does nothing to explain it. Natural Selection cannot move or stir one step, till this double assumption has been made.

The third postulate is partial variation. As a truth of experience, it teaches that the offspring, vegetable or animal, may vary considerably from the parents, but only within certain specific limits. The new theory consists in the assertion or inference, that this variation has no limit of species or kind, and the sole condition that it must be very small in one single generation. This wide extension of the law of variation has no single fact, certainly known, to confirm it, and is opposed by an immense mass of negative evidence. It rests on conjecture and supposition alone. A fourth postulate is the permanence of slight variations, or that they can be transmitted from parents to their offspring, and progressively accumulate through many thousands of generations. A fifth is the permanency, for like intervals, or through immense periods, of physical conditions, so that the same modifications are favourable to life, in each case, throughout a vast number of descents from the origin of the race.

Let us now combine Mr Spencer's definition of Life with Mr Darwin's of Natural Selection, to obtain the exact nature of this newly discovered law, which is said to be a "true cause," and to take in physiology the very same place which is occupied in physics by Newton's discovery of universal gravitation. It will run as follows:

"Every definite combination of heterogeneous changes, simultaneous and successive, in correspondence with external coexistences and sequences, tends to generate other definite combinations of heterogeneous changes, simultaneous and successive, also in correspondence with other and later coexistences and sequences; and the second tend to generate a third, and the third a fourth set of definite combinations of heterogeneous changes, and so on, till the number of these successive sets of definite heterogeneous changes, simultaneous and successive, amounts to many thousands, with the sole condition that each sequent definite combination of heterogeneous changes shall differ from the antecedent definite combination, by a variation very slight in amount, and in character wholly undefined." The formula is pronounced an adequate key to the explication of all the known facts of physiology, or the production of millions on millions of individual plants and animals, each with a structure of wonderful complexity and apparent skill, and ranged in a hundred thousand different species, none of which, or very few at the most, are ever known to overpass the bounds which sever them from each other. On the other hand, the ascription of their origin to a Divine Author, creating parents of each species, with power to propagate after their own kind, is pronounced to be an unintelligible and illegitimate conception, born only in the dull, dark minds of aboriginal men. The extreme hardihood of assertion, with which the true character of the two hypotheses is interchanged, is really gigantic and sublime.

The definition of Life in the Principles of Biology, besides its own difficulties already stated, suggests many others, when we combine it with Mr Darwin's definition of Natural Selection. How can one definite combination of changes, simultaneous and successive, breed another such combination? Why should the two be like each other at all? If alike at all, why are they not perfectly alike? How can a definite combination of changes struggle for existence? How can one definite combination of changes be more favourable than a second to the existence of a third combination, which is in correspondence with later coexistences? How can the aggregate action and product of many sequences of events, ascertained by us, select for itself an incomparably great number of intermediate and transitional links between definite combinations of changes, never seen by us or by mortal eyes, which occurred millions of years before there were human ears to hear of them, or human eyes to observe them? All these unsolved enigmas lie around the roots of this new theory, which pretends to account for the birth and growth of the universe without a wise and intelligent Creator.

The theory may be submitted to a double test. As a mere hypothesis, will it account for the facts it professes to explain? And again, are its conjectural facts, the intermediate links and forms of life which it requires, consistent with the known and certain facts of science, or excluded by them?

First of all, Natural Selection plainly cannot account

for the derivation of any one existing species from any other such species. For both of them do exist, and, so far as we know, have coexisted for more than four thousand years. The lower species must thus have been as favourable as the higher to self-preservation and continuance. There can have been no killing off of less favoured varieties, and survival of more favourable varieties, removed by one step towards the second type. For the first type exists side by side with the second, and has proved itself equally successful in the great struggle for life. Hence the motive power is entirely wanting, on which the whole efficacy of the principle is assumed to depend.

Again, Natural Selection can never increase the number of original species. It requires, as its very condition of working power, that a less happy species should perish and disappear, that a happier and better may be born. Each step of the progress implies a dying out of more imperfect forms that are successively left behind and expire. Grant all its assumptions, and it might explain how a thousand species, ill constructed, like badly armed soldiers, may be changed into a thousand others, better equipped for the struggle with death, their common foe. But it cannot explain how or why one species should vary laterally into a hundred forms.

Thirdly, Natural Selection, by the very definition, can effect no change, unless each modified variety, in the long succession, is more healthy, more favourable to conservation and propagation, than the one which it succeeds. Let A become Z by natural selection, passing through B, C, D, &c. in the course of this progress. Then B must replace A, simply because it is more favourable to the birth and growth of new individuals, and C must replace B for the like reason. A single decline or regression in healthiness, using that word in a large sense, would be fatal to the whole series of changes. If any one intermediate link were less favourable, less fit to sustain the life of the race, than the one before it, the gulf between the extremes could never be passed.

Natural Selection, again, can lead upward to no improved species, unless the improved offspring, in every stage, are somewhat more numerous than their parents. Suppose the number barely equal, and the actual species so numerous as to include ten thousand millions of individuals. One-tenth at least, we may suppose, will die or be killed before the age of reproduction. In this case twenty descents would reduce the number tenfold, and in two hundred descents the number would be reduced to one, and its continuance would thus be impossible. But the hypothesis requires in every change to a new species, at the lowest, some hundreds of generations. Hence no such change is conceivable, except under the condition named above.

Fifthly, Natural Selection can account for no change affecting a single organ or external feature, apart from the rest, but only for such as imply increased healthiness, or greater resistance of the plant or animal, as a whole, to the inroads of death. If a race of dogs, for instance, run faster by having longer legs than their fellows, there can be no tendency to form a new species with this character, if these longer legs occasion greater weakness, and make the animal more liable to fatigue, exhaustion, and death. Natural Selection cannot propagate a special colour in birds, if either their fertility or average strength be lessened at the same time. It cannot account for any tendency to leave one type and approach another, unless the type forsaken be less healthy in its mean form than at one or other of its extreme limits. Thus, if rabbits varied in the direction of longer legs and swifter running, at the price of a lessened power and skill in burrowing, and were thus, on the whole, more exposed to danger than before, the fact that greater speed is a protection, other things being the same, could never help them a single step, by natural selection, to the Darwinian metamorphosis into hares.

Natural Selection, again, apart from creative design, can never account for ascent to higher forms of life, without descent to lower. Variation, with no guide or aim, must be supposed to take place, almost equally, in all directions. Some of the offspring will be larger, others smaller than the average, and each organ either smaller or larger, when it varies from the normal size, or from the standard of its parents. But ascent is usually harder than descent, and in this case also it must surely be easier to fall than to rise. No variation can issue, as we have seen, in an improved species, unless at each stage the variant offspring are somewhat more numerous than the parents. But if some rise above, as many or more will probably sink below, the parental standard of vitality. It seems thus impossible that a vast animated system, even under the assumptions most favourable to the doctrine, could ever be produced, in ascending from the inferior limit, by natural selection in any form.

Again, this principle could never give rise to any new and variant species, unless it were supplemented by a doctrine of caste, ranging through the whole extent of the animal and vegetable world. There must, at each step, be a virtual prohibition of marriage between variant and non-variant individuals. In the absence of such a law, the first variation must melt away and disappear in the second and third generations. The vital capital, acquired by some favoured individuals, would melt into the common stock, and serve merely to repair the loss involved by the pauperism and bankruptcy of others. The notion, however, that insects, infinitesimally improved, would only consent to pair with each other, is evidently absurd.

Once more, Natural Selection, if it could effect any changes at all, must tend to evolve prolific species from the more sterile, not the sterile from the more prolific. In any species fecundity is the most direct and powerful cause of its continuance and increase. Where it exists in a high degree, it needs stern conditions, such as dearth of food, or the spread of a rival species that preys upon it, to hinder excessive increase of numbers. Thus a slight increase of mean fertility may serve to counterbalance and outweigh secondary inferiority of several kinds. A variety, quite sterile, must be extinct in one single generation. The less fertile, compared with the more fertile, must become extinct also, unless this contrast is balanced by greater healthiness in the sterile class. But however possible this may be in special cases, on the large scale it is incredible, and almost impossible. We cannot, without a miracle, assume that, in ten thousand species, a decrease of productive power and superior healthiness and increase of vital energy go together.

It is plain that the higher quadrupeds are less fertile than the lower, the elephant than the ox, the ox than the sheep, the sheep than the rabbit, the eagle or ostrich than the dove or the swallow. Mammals, as a class, are less prolific than fishes, and even fishes than insects. The conditions of healthy existence are more complex and various, as we mount higher in the scale of animated being. The inference seems clear and unavoidable. Natural

Selection, the principle of survival of the fittest, can never account for the production of insects from animalcules, fishes from insects, birds or beasts from fishes, and among birds and beasts, of the higher kinds from the lower. It would then have to reverse completely that law of variation, on which its mighty efficacy is supposed to depend.

In the ninth place, Natural Selection, even when the principle of variation without limit is allowed, could never give birth to any species, which merely maintains its own place under a system of mutual checks and limitations.

Herbivorous animals are limited, not only by the supply of the kind of food which they require, but by the carnivorous instincts of other species. These, in turn, are limited by the amount of prey which they require, and the difficulty of obtaining it. The theory requires us to believe that every species has reached its actual form through ten thousand variations, each more favourable to its own existence than the one before it, less favourable than that which follows. Now the best of all barely sustains itself, or does little more than sustain itself at a constant level, in the struggle for life. Yet the theory requires us to believe that, its ten thousand ancestral varieties, each as we go backward less and less fitted for the strife, have survived and flourished through myriads or millions of years. This seems plainly impossible. Under the conditions which are essential to the theory, every transforming species, with only a very few possible exceptions, must have perished long ago.

Lastly, Natural Selection can effect no birth of new species, unless the long ages it requires for every transformation were free from all physical changes, either wholly destructive or highly injurious to animal and vegetable life. Its principle is that of the slow, continual accumulation of slight changes, beneficial to the life of the species, and under nearly fixed conditions. But once let the conditions alter, and all the previous steps will be lost. The species, in adapting itself to the old conditions of being, may have become less adapted than before to the new ones. The great army of life is either entirely disbanded, or has to be completely re-organized. But there can be no act of parliament to refund the regulation prices, which every species may have paid down, in many steps of change, for its own promotion. The aristocracy of the burgher class, advanced one or two thousand steps towards the ten thousand required for a complete patent of animal nobility, may be swept away in a moment by a flood, a volcanic outbreak, a stifling cometary mist. Only a low proletariat of animalcules and zoophytes might then survive, to work their toilsome way upward, through fresh millions of years, to the lofty mountain-tops of a vertebrate existence.

The main expedient, on which the theory relies, to make the transition from one species to another not wholly incredible, is to divide it into infinitesimal parts, and spread it over thousands or ten thousands of generations. But the difficulty, thus lessened on one side, is increased on the other. The series of changes is thus almost certain to be broken, and its imaginary object frustrated, by fatal and destructive changes in the physical state of the earth in the long course of ten thousands or even millions of years.

Natural Selection, therefore, even as an hypothesis, is loaded with nine or ten conditions, which make it wholly unfit to discharge the gigantic task assigned it, and to account for the origin and continuance of hundreds of thousands of distinct types of animal or vegetable life, each perpetuating itself in successive generations, without the presence or work of any intelligent Mind. It would explain the vast multitude of known and experienced facts, by inventing conjectural facts ten thousand times more numerous, and it still leaves the real phenomena, and its own conjectural additions, equally unexplained. It cannot derive one existing species from another. There is no survival of the fittest, where both alike survive, and the motive power to educe one from the other is thus wholly absent. Bears cannot have turned themselves into whales, by fishing in the water for insects through successive generations; for the conditions of existence have been equally favourable to both for many thousand years. The development of rabbits into hares, penguins into ducks or albatrosses, mice into bats, titmice into shrikes, deer into long-necked giraffes, and chimpanzees or gorillas into men, changes all suggested by Mr Darwin or his disciples, are excluded, to say nothing of other reasons, by this condition alone. An able writer has observed on the same hypothesis, when applied to progress in different kinds of bees;—"Each bee of every variety gets on very well in its line of life. The Mellipona does not fail, and as far as we know, wants nothing, and the humble-bee prospers everywhere. Where, then, was the struggle for existence, or why was it necessary to concoct the new order of architects?" The humble-bee in short, cannot have developed into the hive-bee by the gradual accumulation of desires to construct hexagonal cells by trying to sweep equal imaginary spheres, Mr Darwin's suggestion, nor by the like accumulation of instinctive desires in fertile males and females, to produce eggs breeding sterile females, to secure greater advantages, in the struggle for life, of bee commonwealths long after to come into being.

The ten thousand species of vertebrate animals, by the theory, must have had a hundred millions of preceding varieties, their own forerunners, in the ten thousand stages needed for their origin from some primordial form. All these must have been distinct from each other, and from the species now alive. All these hundred million varieties must have been, in the cosmic sense, less healthy, less fit to escape dangers, and preserve their race, than the ten thousand species which now survive at the goal of this long journey. They need to have varied only upward, when downward change is more easy, and the chances seem to be infinite that variations would be both up and down. They must have grown better suited for permanence as they grew less and less fertile, a result most unlikely, and almost impossible. A strict rule of caste must have prevailed among insects, fishes, birds, and beasts. Animals, infinitesimally better suited to survive, must have refused to couple with those infinitesimally less advanced, through ten thousand generations....Each variety, too, needs to have been less favourable to life than its successor. This must have been true of the mice whose wings were growing slowly, till they became bats, or of the cow, horse, or giraffe, whose tails were developing to help them in the struggle for life. "Suppose at the end of a thousand years the tail had grown an inch long, it would avail little for the flagellation of insects, nor can we see how the slowly improving cow or horse could by this slight change be helped in the competition for existence. How did the pre-gazelles and pre-antelopes escape the carnivorous animals, all the time their legs were lengthening, and their powers of speed accumulating? The middle passage, for all transforming animals, must be their struggle for existence, though not in the sense the theory intends."

The name of the theory is deceptive and untrue. It speaks of selection, where there is no power or faculty of choice. There is a verbal retention, to disguise a real exclusion, of mind or intelligence. Nature is said to mean "the aggregate action and product of many laws," and these laws mean only "the sequence of events as ascertained by us." Yet we are told that this aggregate action and product of many laws "can act on every internal organ, and on the whole machinery of life." Man selects only for his own good; but the aggregate of sequences "selects only for the good of the being which she tends." Every selected character is fully examined by Nature, and placed "under well suited conditions of life." If she has to make the beak of a full-grown pigeon very short for the bird's advantage, "the process of modification will be very slow." This choice by an aggregate of sequences, "if it be a true principle, will banish the belief in the continual creation of new organic beings." No difficulty is seen in this choice, where there is no one to choose and nothing is chosen, by sequences of events, ages before the object of the choice is attained, accumulating variations of instinct to any extent, that may be profitable to the vitality or longevity of insects or animals in distant, unborn generations. The Apostolic law of duty, that the fathers should lay up for the children, thus receives a marvellous extension to the lowest orders of life, and to palæozoic ages of time. "I believe," Mr Darwin says, "that the hive-bee has acquired by Natural Selection (that is, by the choice of the aggregate sequences of events ascertained by us), her inimitable architectural powers." If feeble man can do much by his powers of artificial selection, he can see no limit to the beauty and complexity of the co-adaptations, which may be effected, in the long course of time, by the struggle of living things to keep alive, and the dying off of those which, for some reason or other, cannot live!

In these reasonings it has been assumed that thousands of species of plants and animals do exist, and have existed as far back as human observation extends. But if the doctrine of Natural Selection be true, the existence of species at all is an impossibility and illusion. There are individuals in countless numbers, giving birth to other individuals. But the walls of partition, which have always been held to sever species from each other, fall flat, like the walls of Jericho, and wholly disappear. There are no fixed types, no lines of demarcation. The change in four thousand years is so slight, only because that interval is like a moment, compared with the vast ages in which the law of evolution is supposed to have been stealthily and steadily at work. It does not appear incredible, we are told, "that from some low and intermediate form, both animals and plants have been developed, and that all the organized beings which have ever lived upon earth may have descended from some one primordial form."

On this hypothesis it has been remarked by an able anonymous writer, that "this primordial spore must have been the most marvellous of all beings, to have within itself the potential existences of all animals and vegetables that were ever to be; to possess qualities which were ultimately to expand into an elephant, a whale, a palm tree, an eagle, a crab, a butterfly, and a man. Therefore we may anxiously inquire—Whence came it? Who, or what, were its parents? How was it made? How did it acquire its double quality of animal and vegetable? In what way did the first springing commence? The new spore had to struggle with itself. Or perhaps the great Ancestor produced several modified spores, and thus the struggle began among the family. The unimproved were exterminated, and an advanced race began. A race of what? History does not reveal. Then male and female had to be developed. Natural Selection formed the two sexes, invented all the mysteries of reproduction, and set the world going, till the process finished in man." (Darwinian Theory, pp. 165—167.)

The only consistent meaning of the doctrine, when stripped of all metaphors, and exhibited in its naked simplicity, is that there must be a tendency in those individuals to survive, and propagate their like, which are the best adapted to live. Animalcules, whether amœbæ or hydras, multiplied until the world could hold no more. Then they struggled for life with each other. In this conflict the survivors were larger and stronger than those which perished, and developed thus into worms and insects, fishes, birds, quadrupeds, and men. Worms strove to live, or strove that their descendants might live. Thus, by force of instinctive desires, accumulated from age to age, they became bees and butterflies. Insects strove to live, or to grow healthier, and they thus became fishes, and fishes multiplied, and became beasts or birds. The smaller quadrupeds improved themselves, till they became dogs, kine, elephants, or monkeys, and finally apes or monkeys gave birth to men.

The excuse offered for building a theory of the universe on a metaphor, that it is hard to avoid personifying, is very worthless. It proves only that a deep instinct of the mind, if it be violently repressed by falsely pretended science, will assert itself in some other way. If we shut out the living God, the Almighty Creator, and frame a system of the universe from which the thought of his power and wisdom shall be strictly banished, some idol is sure to be set up in his stead, whether by the old heathens, or by physical philosophers in modern days. The old idols were stocks cut from the living tree, or stones cut out from the solid rock, and then shaped and hewn by art and man's device. The modern idol is just as lifeless. It is either the "sequence of events ascertained by us," or a fixed, invariable amount of something called Force, which means by turns motion, attraction, repulsion, oscillation, instinct, sensation, thought. But this blind nothing, "the sequence of events," or this blind force, which pulls and pushes for ever without any purpose or object except perpetual change, is promoted by the new philosophy into a wise, watchful, beneficent Power, presiding over all events, and always on the watch, like a Moses, a Justinian, or an Alfred, to ameliorate the working laws of the universe, and take advantage of all accidents, to improve and benefit remote generations. It is a gaseous and elastic goddess. Under the pressure of a hostile touch it condenses, and becomes a mere metaphor, an abbreviated method, as misty and obscure as brief, of affirming that the animals best fitted for life will live, and the less healthy will die out and expire. But when the pressure is removed, like the genie in the Arabian tale, she assumes her proper shape, and dilates into a Titan goddess once more. Her feet stand on the fossils of the Pre-adamite earth, but her head is among the stars. Her tongue "walketh through the heavens." By her own unaided power she dispenses with God, and creates a universe. She claims, in her excellency, that there is none else beside her, and that she sits as a queen over all the changes of countless worlds.

I have now examined the chief doctrines, which, under the name of First Principles, and the Philosophy of Evolution, are widely current in these days, have been held up by some professed leaders of scientific progress for the passive acceptance of credulous disciples, and are said to bind Nature itself in the bonds of fate to an extent before unknown. I have proved them to be, from first to last, a strange conglomerate of confusion of thought, error, and self-contradiction. The currency they have gained, and the high-sounding names under which they have been promulgated, are a startling sign how easily men are deceived, when they do not like to retain God in their knowledge, and how readily they can adopt empty sophisms as oracles of wisdom, which, when once touched by the hand of strict analysis, burst like bubbles and disappear. The Doctrine of the Unknowable is a lower depth in the scale of intellectual and spiritual darkness than the old Athenian idolatry. The Persistence of Force and the Indestructibility of Motion, when set up to replace the true and living God of the Bible, the Almighty Creator of heaven and earth, will be found, on enquiry, to be still meaner and more worthless than the old heathen idols of wood and stone. One sentence of the word of God, in the song of the heavenly elders, lays the foundation of a philosophy nobler and deeper than all the human counterfeits of these latter days.

Cambridge:
PRINTED BY C. J. CLAY, M.A. & SON,
AT THE UNIVERSITY PRESS.

www.ingramcontent.com/pod-product-compliance
Lightning Source LLC
Chambersburg PA
CBHW021151230426
43667CB00006B/342